Abnormal Pressures while Drilling

Abnormal Pressures while Drilling

Editor

Awdesh Tiwari

Abnormal Pressures while Drilling

Edited by **Awdesh Tiwari**

Printed in 2017

ISBN: 978-1-68117-327-6

Library of Congress Control Number: 2015939240

Contents

vi

Preface

This book is intended mechanisms and causes of abnormal pressure distribution of a detailed study of the geological and geophysical parameters under different genetic mechanisms of response characteristics of geological logging data response characteristics of drilling engineering parameters and drilling parameters to establish a system response characteristics of abnormal pressure while drilling prediction basic theory put forward a series of new pressure cap pressure distribution method based on the pressure transition zone from three aspects of abnormal pressure while drilling conducted in-depth study of prediction methods and establish abnormal pressure while drilling technology series forecasting procedures and evaluation of abnormal pressure while drilling monitoring methods were studied systematically

Editor

A New Method for Predicting the Position of Gas Influx Based on PRP in Drilling Operations

Xiangwei Kong[1], Yuanhua Lin[1,2], and Yijie Qiu[1]

[1]State Key Laboratory of Oil and Gas Reservoir Geology and Exploitation, Southwest Petroleum University, Chengdu, Sichuan 610500, China

[2]CNPC Key Lab for Tubular Goods Engineering, Southwest Petroleum University, Chengdu, Sichuan 610500, China

ABSTRACT

Accurately predicting the position of gas influx not only helps to analyze complex formation structure, but also can provide reference for taking effective measure, such as increasing mud density, increasing back pressure and casing packer, to suppress the gas influx. Predicting the accurate position of gas influx has been one of the urgent difficulties for drilling industry. With full consideration of the important factors such as the virtual mass force, viscous shear force, energy exchange, and narrow resistance, a new method for predicting the position of gas influx has been proposed based on pressure response time calculation. The gas equations of state (EOS), small perturbation theory, and the fourth-order Runge-Kutta method (R-K4) are adopted to solve the model. Also, the pressure response time plate (PRP) which presents the corresponding relationship between position of gas influx and wellhead parameters by several pressure wave response curves calculated by computer programming is given. The results showed that the PRP is unique at different well depth and gas influx rate, and the position of gas influx can be accurately determined by PRP with known wellhead parameters and detected response time. Therefore, without the help of downhole tools, the accurate mathematical method for predicting the position of gas influx is completely feasible.

INTRODUCTION

One of the future trends of petroleum industry is the exploration and development of high pressure, low permeability reservoirs [1]. Drilling-related issues such as excessive mud cost, wellbore ballooning/breathing, kick-detection limitations, difficulty in avoiding gross overbalance conditions, and differentially stuck pipe and resulting well-control issues together contribute to the applying of managed pressure drilling (MPD) technology [2, 3]. Although drilling operations try to avoid the risk of gas influxes in MPD operations, occasionally there are gas influxes for various reasons. Since the subsequent influx of gas displaces drilling mud, it decreased the pressure in the wellbore and makes gas enter even faster [4, 5]. Gas influx occurs whenever the pressure of a gas-bearing formation exceeds the pressure at the bottom of a wellbore [6]. The main reasons for gas influx are this pressure differentia; the pressure differentia is an unexpected form of rise in formation pressure or a decrease in mud hydrostatic pressure. A rise in formation pressure can be due to geological processes that have occurred in the region being drilled. Wells are drilled in regions where oil and gas are trapped, and the same processes that create the hydrocarbons can also produce large pressures. Therefore, it is not uncommon to come across regions of abnormally high formation pressure while drilling. Mud hydrostatic can decrease due to any event that causes the mud column in the hole to drop, such as lost circulation or tripping out while not filling the hole to adequately compensate for the volume of the removed drilling assembly [7]. Surge pressure, low drilling mud density, abnormal formation pressure and so forth all can cause that the formation pressure be higher than annulus pressure during MPD process, and the higher formation pressure can lead to gas influx from formation to annulus. At any operation condition, the negative pressure exists between the annulus and the formation when gas influx occurs. If the gas influx cannot be detected in time and take effective measures, the negative pressure differentia will further increase with the migrate of gas along the annulus from bottomhole to wellhead. It can result in further deterioration of influx which may escalate into a blowout creating severe financial losses, environmental contamination, and potentially loss of human lives. Normally, gas influx occurs in bottomhole or casing shoe. But when drilling in complex formation, gas influx may occur anywhere in the open wellbore. Also, gas influx may occur at any time while wellbore pressure falls below formation pore pressure in a permeable and porous zone containing fluids. For drilling safety reasons, the sooner it can be detected, the better it will be. Accurate determination of gas influx position is more conducive to take effective measure for suppressing gas influx [8, 9].

Since there are more unknowns, predicting of formation parameters and borehole fluid parameters is always difficult in the drilling industry. At present, the predicting methods generally include software base monitoring and hardware techniques with the help of measurements-while-drilling tools. In drilling site, many judgment methods are put into use, such as level monitoring of return drilling mud in drilling fluid pot, DC index method, shale density method, torque gauge method, acoustic time difference method, and pump speed method. DC index method relies on the accurate determination of normal pressure trend line. For lack of reliability of pressure monitoring and drilling parameters before drilling, the DC index method has limitations. Acoustic curve detection of formation pressure based on acoustic time difference principles is used for prediction of the single well drilling area or regional formation pressure and regional formation pressure profile, which is common and effective. Acoustic velocity is relevant to the density of the rock structure, porosity of formation, and buried depth. The basic principle of acoustic time difference method is that the propagation velocity of sound waves is different in gas drilling fluid and drilling fluid. Seismic reflection wave method is widely used in geophysical methods. Seismic wave method to predict the formation pressure is according to seismic wave velocity difference to decide the formation pressure. The basic principle of pump speed method is based on working mud pump. The mud pump can be seen as a surface pressure pulse generator. The pressure pulses generated by piston in pump enter the circulatory system, such as the drill string, downhole, drill bit, the nozzle, and return to the ground along the annulus [10–12]. MWD tool is also an important means for detection of downhole information. In the early 2000s, formation pressure while-drilling tools were introduced. That can obtain formation pressure data, even in highly deviated wells and extended-reach drilling. In earlier research, a numerical solution of the equations that govern unsteady fluid flow is developed by Chen et al. in 2005. The boundary conditions are adjusted for the surface and downhole equipment. The program outputs pressure and flow pulse predictions at any point [13]. In the past many years, this LWD technology has evolved with the addition of downhole fluid sampling and fluid analysis. LWD sampling and testing are now performed in challenging environments that cannot be performed with wire line tools such as horizontal or highly deviated wells [14]. The new generation of LWD and MWD tools was specifically designed by Radzinski and LWD in 2004 for such hostile environments, transmitting real-time directional information, gamma ray, bore and annular pressure, vibration data, resistivity, neutron porosity, and density measurements [15]. Chia quantified a significant improvement to standard MWD surveyed position uncertainty using actual survey data from drilling assemblies used in more than 120 runs in over 35 different wells in 2004. The use of multistation analysis and the subsequent reduction in wellbore position uncertainty can reduce overall surveying and drilling costs for the well, removing the need for correction runs and allowing for penetration of smaller targets than previously possible with standard MWD surveying [16]. Wang demonstrated that the application of MWD is not limited to streamer data but can also be extended to ocean bottom seismic (OBS) data. For OBS data, MWD can remove water-layer-related multiples and receiver ghost in one step [17]. An MWD data transmission system and method were provided for determining and transmitting the environmental properties of the downhole borehole assembly (BHA) to surface data receivers via mud pulse telemetry, EM telemetry, or both mud pulse telemetry and EM telemetry based on one or more determined properties of the downhole environment by Young in 2012 [18]. Construction and a field testing of a prototype that can automatically record data while drilling from caving and influx flow were analyzed in real time. The prototype is used to identify situations in which influx and caving flows are high enough to cause instability of the drilled well in real time. Geraud et al. combined different services, such as the BHA which included a rotary steerable system (RSS), measurement and telemetry service, logging-while-drilling (LWD) magnetic resonance service, multifunction petrophysics platform, formation pressure service, and sonic and seismic services in 2013. Measurements from these services are integrated and used for real-time drilling parameter optimization and formation evaluation [19].

Though some efforts have been made, predicting the position of gas influx still depended on measurement while drilling (MWD). In the past researches, the influencing factors for the position of gas influx are simulated and analyzed with the MWD; however, the variation of gas void at different depth of wellbore is not considered [20–22]. The current researches are limited in their assumption and neglect of the flow pattern translation and interphase forces along the annulus. Up to now, no mathematical method to predict position of gas influx in annulus with variation of gas void, flow pattern, temperature, and BP during MPD operations has been derived. The research in this paper will hopefully solve the present puzzle for predicting the position of gas influx occurring in drilling operations. In this paper, the new method for predicting the position of gas influx is proposed based on acoustic time difference method. So, the determination of pressure wave velocity is the key of this method. Since the 1940s, many experimental and theoretical studies for pressure wave velocity have been performed. Experimental tests are conducted to inspect the contributions of fluctuation and flow characteristics on pressure wave. Pressure wave is still worth continuing an in-depth study today [23–29]. In drilling industry, Wang and Zhang studied the pressure pulsation in mud and set up a model for calculating the amplitude of pressure pulsation when pressure wave transmits in drilling-fluid channel especially drilling hose with different inside diameter [30]. Lin et al. study the wave velocity for the transmission of pressure disturbance in the two-phase drilling fluid in the form of a pressure wave in annulus during MPD operations in 2013 [31].

The purpose of this paper is to describe a new method based on pressure response time plate (PRP) for predicting the position of gas influx in the two-phase flow in annulus during MPD operations. In addition to pressure, temperature, and void fraction in the annulus, the compressibility of the gas phase, the virtual mass force, and the changes of interface in two-phase are also taken into consideration. By introducing the pressure gradient equations in MPD operations, gas-liquid two-fluid model, the gas equations of state (EOS), and small perturbation theory, the method for predicting the position of gas influx in gas and drilling mud in annulus is developed. The method can be used to predict the position of gas influx at different influx rate, applied back pressure, and well depth with a full consideration of drilling mud compressibility and interphase forces.

THE MATHEMATICAL MODEL

The drilling system described is an enclosed system (Figure 1). The drilling mud is pumped from surface storage, down the drill pipe. Returns from the wellbore annulus travel back through surface processing. The key equipment include pressure sensor, choke, and gas-liquid flow meter as follows.

- Pressure sensor: a pressure sensor is used to measure surface back pressure on the wellhead.
- Electronic valve: the MPD choke manifold provides an adjustable choke system which is used to dynamically control the required BHP by means of applying surface BP.
- Gas-liquid flow meter: a gas-liquid flow meter is used to accurately measure the mass flow rate of fluid exiting the annulus. The ability to measure return flow accurately is essential for the applied-back-pressure.

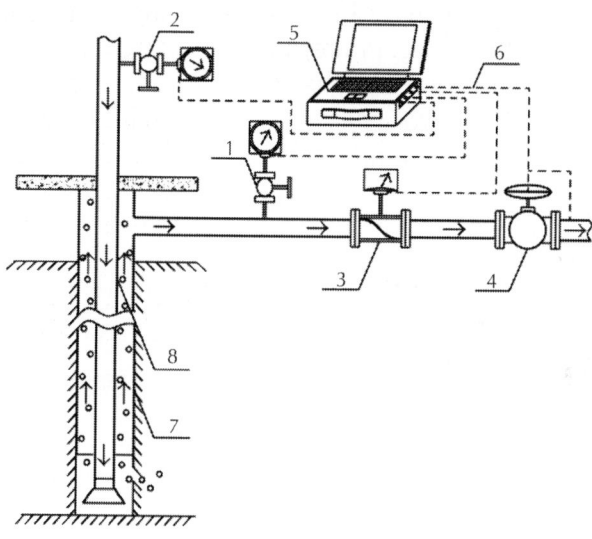

Figure 1: Schematic diagram of pressure response testing ((1) back pressure sensor, (2) stand pipe pressure sensors, (3) gas-liquid flow meter, (4) electronic valve, (5) IPC (industrial personal computer), (6) command lines, (7) casing pipe, and (8) drill pipe).

Both gas-drilling mud flow rate measured by gas-liquid flow meter and the back pressure measured by pressure senor are the initial data for pressure response time calculation in annulus.

During MPD operations, if the gas influx occurs in the bottomhole, the pressure wave velocity will be significantly reduced. This is due to the low density and great compressibility of gas. Gas migrates from the position of gas influx to the wellhead along the annulus. Above gas influx position, fluid is composed of gas and two-phase drilling mud. Below gas influx position, fluid is single-phase drilling mud. When gas migrates to the wellhead, the degree of the electronic valve is increased to suppress gas influx occurrence. Adjustment of throttle valve to increase back pressure and suppress gas influx will generating pressure pulse. Pressure pulse propagates from wellhead to bottomhole along the annulus in the form of pressure wave. After arriving at bottomhole, the pressure pulse returns back the wellhead in two different paths, propagating along the drill pipe and propagating along the annulus. The pressure sensor detects the difference of propagation time T_c of the two paths.

Gas influx position accuracy detection relies on the calculation of the pressure response time. Pressure response time is the propagation time of pressure wave from wellhead to bottomhole. The wellbore can be divided into several grids. The algebraic sum of propagation time in every grid is the response time.

Divide the annulus into n discrete grids, and the first grid is at the wellhead ($H_0 = 0_m$, void fraction is ϕ_0, BP is P_0, andwave velocity is c_0). According to the wellhead parameters, the next parameters (ϕ_1, $P_{1,1}$, c_1) of the grid can be calculated by Runge-Kutta method, followed by the ith grid parameters (ϕ_i, P_i, H_i, c_i) which are obtained. According to the parameters of the ith grid, pressure response time on the corresponding grid can be obtained.

In MPD operations, calculation equation of pressure wave response time based on the wave velocity and well depth is expressed as follows:

$$T\left(H_i\right) = \sum_i \frac{H_i}{c_i}, \quad (i \leq n),$$

(1)

where $T(H_i)$ is the pressure response time on nod i (s); H_i is wellbore length on nod i (m); c_i is wave velocity on nod i (m/s).

If the drilling fluid is composed of gas and two-phase drilling mud in annulus, the wave velocity can be expressed as

$$c_i = c_{gli}\left(P_i, T_i, \phi_i, w_i, Lg_i\right),$$

(2)

where c_{gli} is wave velocity in gas-drilling mud phase on nod i (m/s); P is pressure (MPa);$T$is temperature (K); ϕ is gas void fraction; w is angle frequency (Hz); Lg is flow pattern.

If the drilling fluid in drilling pipe is single-phase drilling mud, the wave velocity can be expressed as

$$c_i = c_1,$$

(3)

where C_i is wave velocity in single-phase drilling mud (m/s).

Above the gas influx position (nod i), the fluid in annulus is gas and drilling mud two-phase mixture. Pressure response time of pressure pulse in the annulus above the nod i can be expressed as T_1. Consider

$$T_1 = \sum_i \frac{H_i}{c_i\left(P, T, \phi, w, Lg\right)}, \quad i \le n.$$

(4)

Above the gas influx position (nod i), the fluid in drill pipe is single-phase drilling mud. Pressure response time in drill pipe is expressed as T_2. Consider

$$T_2 = \sum_i \frac{H_i}{c_1}, \quad i \le n.$$

(5)

Response time difference ΔT is obtained by calculation.

$$\Delta T = T_1 - T_2.$$

(6)

Response time difference T_c is obtained by the detection of two pressure sensor. The precision is defined as δ. Consider

$$\left|T_c - \Delta T\right| < \delta,$$

(7)

where T_c is response time difference between the two paths detected by pressure sensor; ΔT is the calculated time difference (s); δ is computational accuracy (s).

The gas influx position can be determined by

$$H = \sum_i H_i, \quad i \le n,$$

(8)

where H is the length of wellbore above the gas influx position (m). If the influx occurs at the bottomhole, H amounts to the depth of the well.

Furthermore, each position of gas influx and gas influx rate corresponds to a pressure wave response curve. The PRP presents the corresponding relationship between position of gas influx and wellhead parameters by several pressure wave response curves calculated by the computer programming. With known gas influx rate and pressure wave response time, the position of gas influx can be accurately determined on the basis of the PRP.

GOVERNING EQUATIONS

In the following part of this section, the calculation equations of pressure wave velocity and flow parameters very well depths are given.

Wave Velocity in Gas and Drilling Mud Two-Phase Fluid

The continuous equation for gas phase can be expressed as follows:

$$\frac{\partial}{\partial t}\left(\phi\rho_g\right) + \frac{\partial}{\partial s}\left(\phi\rho_g u_g\right) = 0,$$

(9)

where ρ_g is gas density (kg/m³); u_g is gas flow velocity (m/s).

The continuous equation for drilling mud phase can be expressed as follows:

$$\frac{\partial}{\partial t}\left[(1-\phi)\,\rho_l\right] + \frac{\partial}{\partial s}\left[(1-\phi)\,\rho_l u_l\right] = 0,$$

(10)

Where ρ_l is drilling mud density (kg/m³); u_l is drilling mud flow velocity (m/s).

The momentum conservation equation for gas flow can be expressed as follows:

$$\frac{\partial}{\partial t}\left(\phi\rho_g u_g\right) + \frac{\partial}{\partial s}\left(\phi\rho_g u_g^2\right) = -\frac{\partial}{\partial s}\left(\phi P_g\right) + \frac{\partial}{\partial s}$$

(11)

where P_g is gas phase pressure (N/m²); τ_g^{fr} is shear stresses of gas interface (N/m²); τ_g^{Re} is Reynolds stress of gas interface (N/m²); M_{gi} is momentum transfer in gas interface (N/m²); τ_g is shear stresses of gas interface (N/m), and   is annulus effective diameter (m).

The momentum conservation equation for drilling mud flow can be expressed as follows:

$$\frac{\partial}{\partial t}\left(\phi_l\rho_l u_l\right) + \frac{\partial}{\partial s}\left(\phi_l\rho_l u_l^2\right) = -\frac{\partial}{\partial s}\left(\phi_l\rho_l\right)$$

$$+ \frac{\partial}{\partial s}\left[\phi_l\left(\tau_l^{fr} + \tau_l^{Re}\right)\right] + M_{li} - 4\frac{\tau_l}{D},$$

(12)

where , P_l gas phase pressure (N/m²); τ_l^{fr} is shear stresses of drilling mud interface (N/m²); τ_l^{Re} is Reynolds stress of drilling mud interface (N/m³); τ_l is shear stresses of drilling mud along well wall (N/m²); M_{li} is momentum transfer in drilling mud interface (N/m²).

Pressure gradient within the annulus consists of weight component, acceleration component, and friction forces component. Based on the theory of two-phase flow, equation used to calculate pressure gradient of drilling fluid can be written as

$$\frac{dP}{dH} = \left(\frac{dP}{dH}\right)_e + \left(\frac{dP}{dL}\right)_f + \left(\frac{dP}{dH}\right)_{ac},$$
(13)

where $(dP/dH)_e$ is weight component; $(dP/dH)_{ac}$ is acceleration component; $(dP/dH)_f$ is friction forces component.

The total pressure drop gradient is the sum of pressure drop gradients due to potential energy change, kinetic energy, and frictional loss. By simplifying, (13) used to calculate pressure gradient of gas drilling mud two-phase flow within the wellbore can be written as

$$\frac{dP}{ds} = \rho_m g \sin\theta - \frac{\tau_w \pi D}{A} - \rho_m v_m \frac{dv_m}{ds},$$
(14)

where τ_w is frictional pressure of pipe wall (N/m); ρ_0 is the average density (kg/m³).

By differential treatment of the two-fluid model, (9)–(12) is converted to vector by aid of Taylor formula. The small perturbation theory is also applied to the solution of wave velocity model. According to the solvable condition of the homogenous linear equations that the determinant of the equations is zero, the equation of pressure wave can be expressed in the following form:

$$
\begin{vmatrix}
\left(F_a + \rho\phi u_l\frac{u_l^2}{c_l^2}\right)w & \frac{\phi}{c_l^2}\left|1 + c_i\phi_l\right|\frac{u_l^2}{c_l^2}w & \left[\phi_l g^, k + 2c_p\phi\phi_l\frac{u_g}{c_l^2}w\right] & 2\frac{2}{2}c_p\phi\phi_l u_l\frac{u_g}{c_l^2}w \\
\phi u^m & 1 - \phi_l & 0 & -k(1-\phi)n \\
\rho u_l^2 k\left(\phi_{l_p} + c_v + c_l + c_{m3}\right) & \phi k\left[1 - \phi_l\frac{\rho_l^,u_l^2}{c_l^2} + c_l\frac{u_l^2}{c_l^2}\right] & \phi\left(F_a + c_{m1}u_l^2\right)w & -C_m\phi_l\eta w + i\left(\frac{3}{4}\frac{C_D}{r}\rho_g\phi u_s + \frac{4}{D}f_{m1}\rho_l u_l^2\right) \\
& & -i\left(\frac{3}{4}\frac{C_D}{r}\rho_g\phi u_s + \frac{4}{D}f_{m1}\rho_l u_l^2\right) & \\
\rho u_l^2 k\{\phi_{l_p} + 2c_v + c_{m3}\} & k\left(\phi_l + c_l\phi\frac{u_l^2}{c_l^2}\right) & -C_m\phi_l\eta w + i\left(\frac{3}{4}\frac{C_D}{r}\rho\phi u_s\right) & \rho_l|\phi_l + \phi^,c_m|w \\
& & & i\left(\frac{3}{4}\frac{C_D}{r}\rho\phi u_s + \frac{4}{D}f_l\rho u_l^2\right)
\end{vmatrix} = 0,
$$
(15)

where $c_i = 0.3$; $c_p = 0.25$; $c_{m2} = 0.1$; $c_r = 0.2$; u_s is slip velocity (m/s); f_l is shear stresses coefficient of drilling mud interface; CD is the coefficient of drag force; C_{vm} is the coefficient of virtual mass force;

f_{gw} is shear stresses coefficient of drilling mud interface; w is angle frequency (Hz); r is average diameter of the bubble (m).

The real value of wave number is determined the pressure wave velocity, and pressure wave velocity in the two-phase flow is defined by

$$c_{gli}(P, T, \phi, w, Lg) = \frac{\left| w/R^+(k) - w/R^-(k) \right|}{2},$$

(16)

where k is wave number; $\text{Re}(k)$ is the real part.

When $= 0$, the $_{li}$ can be expressed as

$$c_{li} = \frac{1}{\sqrt{\rho_l\left((1/\rho_l)(d\rho_l/dP) + (1/A)(dA/dP)\right)}}.$$

(17)

Flow Pattern Analysis

Based on the analysis of flow characteristics in the closed drilling system, it can be safely assumed that the flow pattern in wellbore is either bubble or slug flow. The flow pattern transition criteria for bubbly flow and slug flow given by Orkiszewski et al. are used to judge the flow pattern in the gas-drilling mud two-phase flow [32, 33].

For bubbly flow, the empirical relations can be expressed as follows:

$$\frac{q_g}{q_m} < L_b.$$

(18)

For slug flow, the empirical relations can be expressed as follows:

$$\frac{Q_g}{Q_m} > L_b, \qquad N_{gv} < L_s,$$

(19)

Where Q_g is volume flow rate for gas (m³/s); Q_m is mixture volumetric flow rate for gas-drilling mud (m³/s).

The dimensionless number L_b is defined by

$$L_b = 1.071 - \frac{0.7277v_m^2}{D},$$

(20)

where V_m is mixture flow velocity for gas and drilling mud (m/s). The dimensionless number L_c is defined by

$$L_s = 50 + 36N_{gv}\frac{Q_l}{Q_g},$$

(21)

where Q_l is volume flow rate for the drilling mud (m²/s).
The N_{gv} can be defined by

$$N_{gv} = v_s\left(\frac{\rho_l}{(g\sigma_s)}\right)^{0.25},$$

(22)

where g is acceleration due to gravity (m²/s); σ_s is surface tension (N/m²).

The mixture density of two-phase flow is

$$\rho_m = \phi_l\rho_l + \phi\rho_g,$$

(23)

where ρ_m is gas and drilling density (kg/m³); ϕ is gas void fraction; ϕ_l is drilling mud holdup.

Drilling mud holdup can be expressed as follows:

$$\phi_l = 1 - \phi.$$

(24)

Bubble Flow

Gas void fraction for bubble flow is

$$\phi_g = \frac{v_{sg}}{S_g\left(v_{sg} + v_{sl}\right) + v_{gr}},$$

(25)

where V_{sg} is superficial gas velocity (m/s); V_{sl} is superficial drilling mud velocity (m/s).

The value of the distribution factor S can be described as

$$S_g = 1.20 + 0.371 \left(\frac{D_i}{D_o} \right),$$

(26)

where D_i is diameter of the inner pipe (m); D_o is diameter of the outer pipe (m).

Superficial gas velocity v_{gr} can be described as

$$v_{gr} = 1.53 \left[\frac{g\sigma_s \left(\rho_l - \rho_g \right)}{\rho_l^2} \right]^{0.25}.$$

(27)

Slug Flow

The distribution factor S_g for slug flow can be described as

$$S_g = 1.182 + 0.9 \left(\frac{D_i}{D_o} \right).$$

(28)

For slug flow, the slip velocity v_{gr} can be calculated as

$$v_{gr} = \left(0.35 + 0.1 \frac{D_i}{D_o} \right) \left[\frac{g D_o (\rho_l - \rho_g)}{\rho_l} \right]^{0.5}.$$

(29)

Physical Equations

Equations of State for Drilling Mud

Under $T \leq 130°C$, drilling mud density was measured by *Xypwyдoe* in different temperature, and the empirical formula is expressed as follows [25]:

$$\rho_{PT} = 100\rho_0 \left(1 + 4 \times 10^{-10} P_l - 4 \times 10^{-5}T - 3 \times 10^{-6}T^2 \right)$$

(30)

Here, ρ_0 is density under standard atmospheric pressure (kg/m³); P_l is pressure of drilling mud (MPa); T is temperature (K).

Equations of State for Gas

State of acidic gas is governed by Redlich-Kwong equation:

$$P = \frac{RT}{V - b} - \frac{a}{T^{0.5}V\,(V + b)},$$

(31)

where V is gas volume; R is gas constant.

Both a and b parameters can be defined by

$$a = \left(\sum y_i a_i^{0.5}\right)^2, \qquad b = \sum y_i b_i.$$

(32)

Here,

$$a = \left(\sum y_i a_i^{0.5}\right)^2, \qquad b = \sum y_i b_i.$$

(33)

where $;\Omega_a = 0.42748$; $\Omega_b = 0.08664$.

SOLUTION OF THE MODEL

Obtaining the analytical solution of the mathematical models concerned with flow pattern, void fraction, characteristic parameters, and pressure drop gradient are generally impossible for two-phase flow. In this paper, the Runge-Kutta method (R-K4) is used to discretize the theoretical model.

We can obtain pressure, temperature, gas velocity, drilling mud velocity, and void fraction at different annulus depth by R-K4. The solution of pressure drop gradient equation (14) can be seen as an initial value problem of the ordinary differential equation:

$$\frac{dP}{ds} = F\,(s, P),$$

$$p\,(s_0) = P_0.$$

(34)

With the initial value $(z_0,\ p_0)$ and the function (z, p), (35) can be obtained as follows:

$$k_1 = F(s_0, P_0),$$

$$k_2 = F\left(s_0 + \frac{h}{2}, P_0 + \frac{h}{2}k_1\right),$$

$$k_3 = F\left(s_0 + \frac{h}{2}, P_0 + \frac{h}{2}k_2\right),$$

$$k_4 = F(s_0 + h, P_0 + hk_3), \tag{35}$$

where h is the step of well depth (m).

The pressure on the nod $i = i + 1$ can be obtained by

$$P_{i+1} = P_i + \Delta P = P_i + \frac{h}{6}(k_1 + 2k_2 + 2k_3 + k_4). \tag{36}$$

In the present work, the mathematical model and pressure wave velocity calculation model are solved by computer programming on VC++ (Version 2010). The solution procedure for the gas influx position is shown in Figure 2. At initial time, the wellhead back pressure, wellhead temperature, wellbore structure, well depth, and gas and drilling mud properties, and so forth are known. On node i, the pressure wave velocity, pressure gradient, temperature, and the void fraction can be obtained by adopting R-K4. Then, the pressure wave response time T_1, T_2 is calculated based on the calculated parameters, compared with the response time difference detected by sensor. The process is repeated until meeting the accuracy requirement. As the accuracy requirement is met at node i, the gas influx occurs at node i. Finally, the distance from the wellhead H can be obtained.

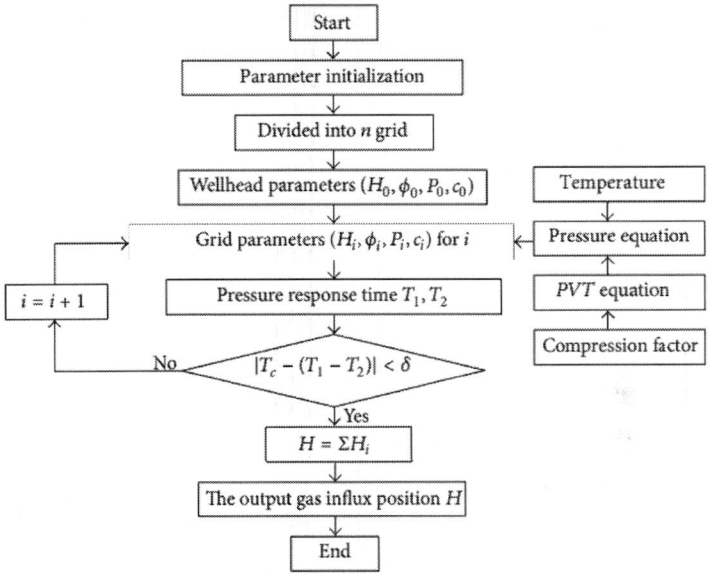

Figure 2: Solution procedure for gas influx position in MPD operations.

ANALYSIS AND DISCUSSION

Both gas and drilling mud flow rate measured by gas-liquid flow meter and the BP measured by pressure senor are the initial data for annulus pressure calculation. The experiment well MF6# used for calculation is a gas well in Sichuan Chengdu Region, Southwest China, and the response time test was conducted on May 23, 2013. The wellbore structure, well design parameters (depths and diameters), gas-drilling mud properties (density and viscosity), and operational conditions of calculation well are displayed in Table 1. The length of well is 4000 m, which is divided into 1000 grids. The length of each grid is 10 m in the calculation. Figure 3(a) shows the experimental equipment in MPD field. A dynamic pressure senor is used to measure pressure disturbance time at the wellhead and pressure disturbance return time to verify the pressure response time.

Table 1: Basic parameters

Parameter	Value
The length of drill collar (m)	200
The length of drill pipe (m)	3800
Diameter of bit (m)	0.2159
Outside diameter of drill pipe (m)	0.127
Inside diameter of drill pipe (m)	0.1086
Outside diameter of drill collar (m)	0.1778
Inside diameter of drill collar (m)	0.078
Drill pipe roughness (m)	0.0154
Flow rate for mud pump (m³/s)	0.037
String elastic modulus (Pa)	2.07×10^{11}
Drilling mud density (kg/m³)	1460
String Poisson's ratio	0.3
Drilling fluid compressibility (1/kPa)	5.7×10^{-8}
Surface temperature (°C)	25
The ground atmospheric pressure (MPa)	0.101
The wall roughness (m)	0.1

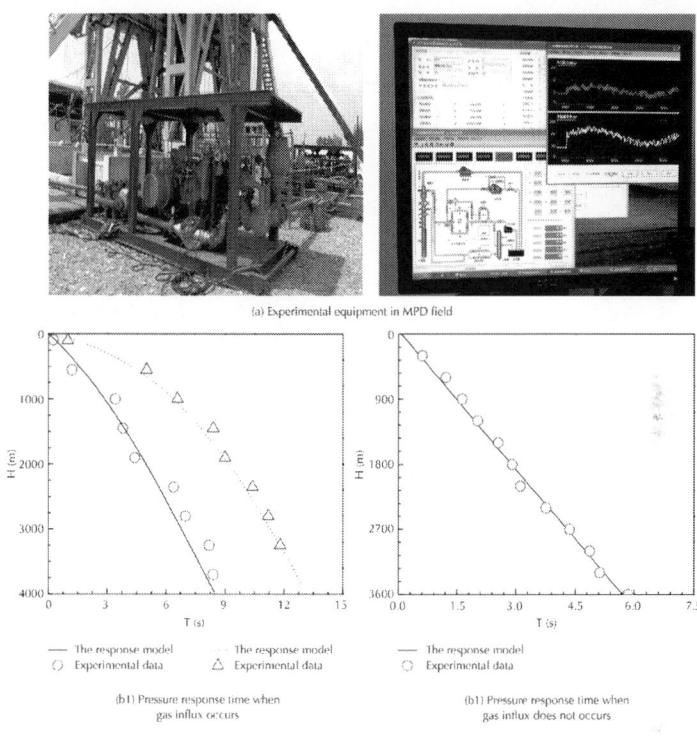

Figure 3: Experimental verification in comparison with field data.

The calculated pressure response time plate is given in Figures 3(b1) and 3(b2). Individually, the two plates present the corresponding relationship between response time and position of gas influx when gas influx occurs or not. Experimental results show that the pressure response time has good consistency with the experimental data. The computer programming can be installed in IPC in real-time in drilling site, and the precision can meet the engineering requirements.

The calculated PRP is unique at different gas influx position and gas influx rate. According to the corresponding principle, the gas influx position can be determined on the basis of the PRP with the known detected response time different when gas influx occurs. Table 2 lists the position of gas influx predicted based on the method in this paper during drilling operations. The key parameters, wave velocity and pressure response time, used for gas influx position predicting are also analyzed in Figures 4–9.

Table 2: Predicted position of gas influx during MPD operations

Receive data from IPC				Test result (m)
BP (MPa)	Drilling mud flow (L/s)	Gas flow (L/s)	Time difference T_c(s)	
0.07	38	23.63	10.25	3081
0.11	37	25.12	11.25	3162
0.15	38	15.29	8.75	2925
0.25	45	5.76	7.25	3286
0.27	41	0	0	No gas influx
0.35	37	0	0	No gas influx
0.51	46	0	0	No gas influx
0.61	39	0	0	No gas influx

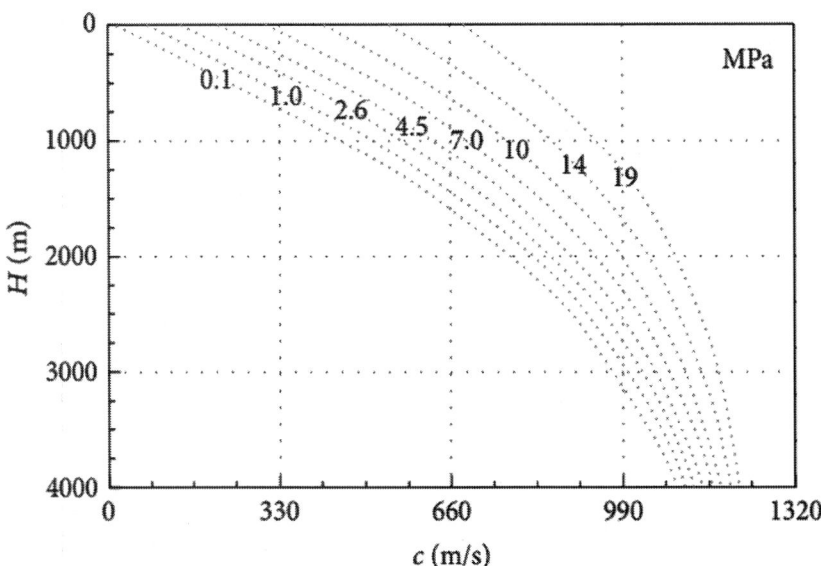

Figure 4: Wave velocity distribution at different BP.

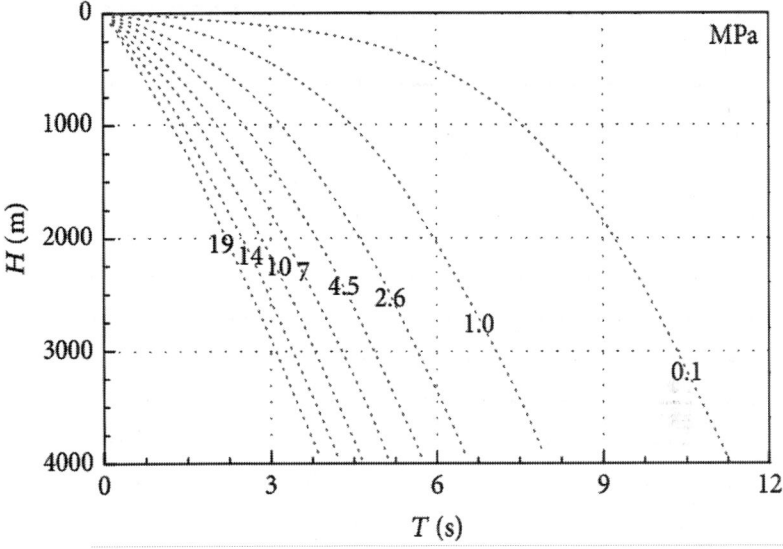

Figure 5: Pressure response time variations at different BP.

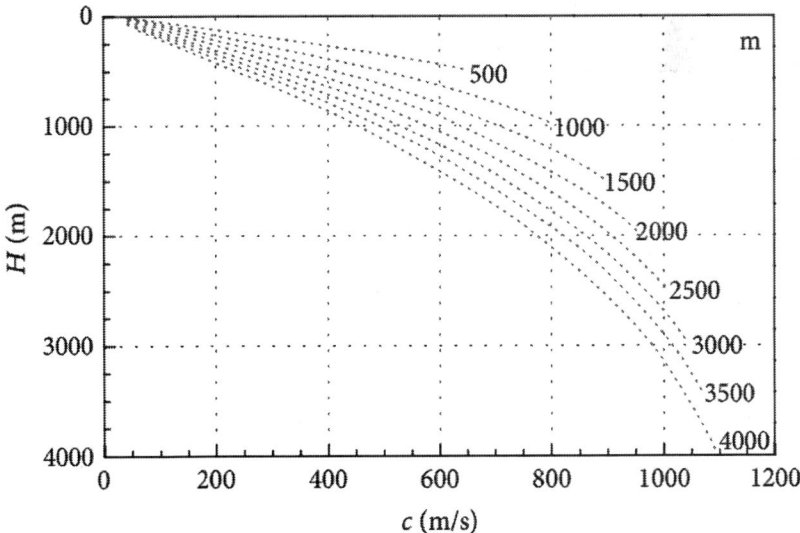

Figure 6: Wave velocity distribution at different well depth.

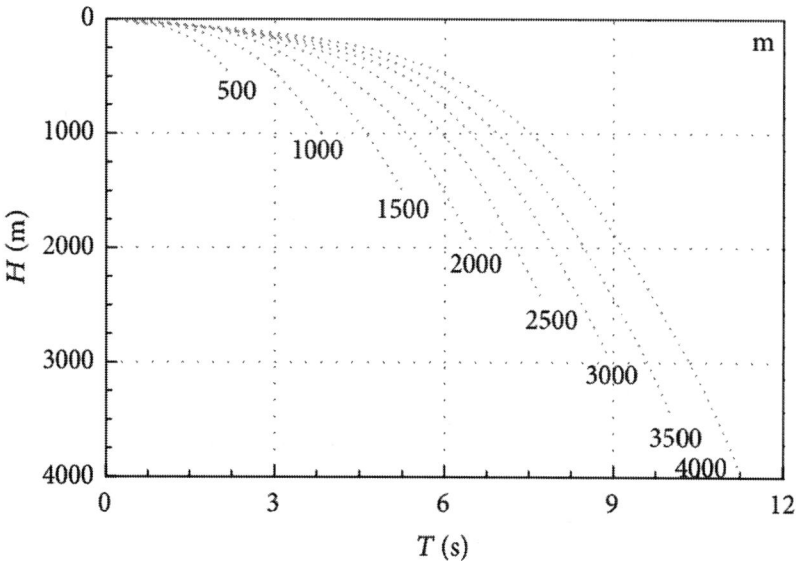

Figure 7: Pressure response time variations at different well depth.

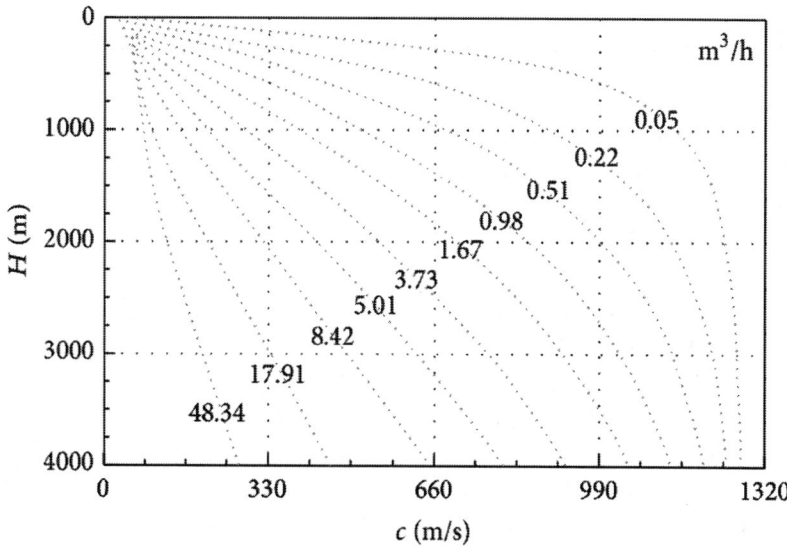

Figure 8: Effect of gas influx rate on the wave velocity.

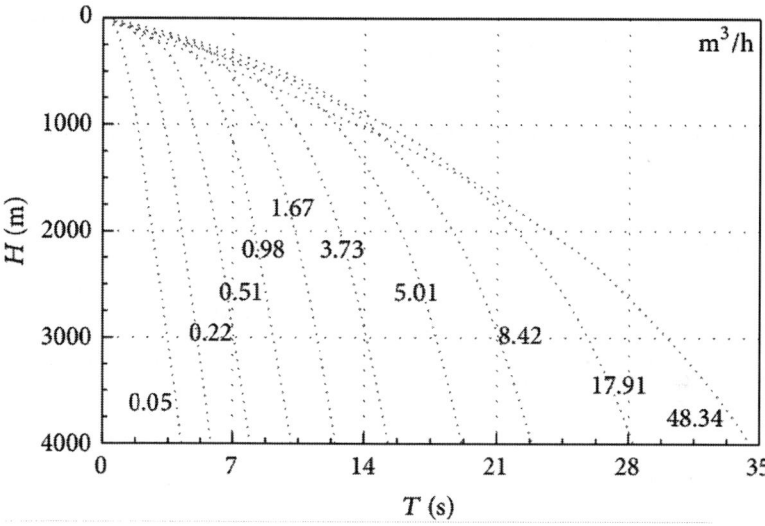

Figure 9: Effect of gas influx rate on the pressure response time.

Effect of BP on PRP

Figures 4 and 5 show the distributions of wave velocity and variations of pressure response time along the flow direction in the annulus when the back pressure at the wellhead is BP = 0.1 MPa, BP = 1.0 MPa, BP = 2.6 MPa, BP = 4.5 MPa, BP = 7.0 MPa, BP = 10.0 MPa, BP = 14.0 MPa, and BP = 19.0 MPa, respectively. It can be seen that the wave velocity significantly decreases along the flow direction in the annulus. Conversely, the pressure response time shows a remarkable increase tendency. This can be explained from the viewpoints of mixture density and compressibility of two-phase fluid and the pressure drop along the flow direction in the wellbore. According to the EOS, if gas invades into the wellbore with a small amount in the bottomhole, the density of the drilling mud has little variation while the compressibility increases obviously, which makes the wave velocity decrease, and the pressure response time shows an increase tendency. Then, the gas migrates from the bottomhole to the wellhead along the annulus with a drop of pressure caused by potential energy change, kinetic energy, and frictional loss, which leads to an increase of pressure response time.

If the wave velocity is increased, resulting in a decrease for pressure responses time along the flow direction.

Effect of Well Depth on PRP

Figure 6 presents the change of pressure wave velocity in the annulus at different well depth. Figure 7 shows the effect of well depth on pressure response time in gas-drilling mud flow. When the gas influx (Q_g = 1.28m³/h) occurs at different well depth (such as H = 500m, H = 1000m, H = 1500m, H = 2000m, H = 2500m, H = 3000m, H = 3500m, and H = 4000 m), gas invades into the wellbore and migrates from the bottom hole to the wellhead along the flow direction. It can be clearly seen from the curves that the wave velocity and pressure response time are varied in real time due to variation of pressure along annulus. With the increase of well depth, both wave velocity and pressure response time are increased. The wave velocity in the two-phase drilling fluid and the distribution of pressure response time at different depth of the annulus will diverge. In conclusion, the wave velocity and pressure response time increase accompanying the increase of the well depth.

Effect of Gas Influx Rate on PRP

Figures 8 and 9 present the variations of wave velocity and pressure response time along the flow direction in the annulus during MPD operations at different gas influx rate (such as Q_g = 0.05m³/h, Q_g = 0.22m³/h, Q_g = 0.51m³/h, Q_g = 0.98m³/h, Q_g = 1.67m³/h, Q_g = 3.73m³/h, Q_g = 5.01m³/h, Q_g = 8.42m³/h, Q_g = 17.91m³/h, and Q_g = 48.34m³/h) in bottom hole. When gas influx occurs in the bottom hole, gas invades into the wellbore and migrates from the bottom hole to the wellhead along the flow direction. It is extremely obvious that the wave velocity and pressure response time first change slightly then sharply change in a comparatively smooth value. The compressibility of the gas is high at wellhead, which results in a change of wave velocity and pressure response time. Since the compressible component increases with the increase of gas influx rate, the compressibility of the gas and drilling mud two-phase fluid is improved, so the variations of wave velocity and pressure response time become more prominent. Under the high bottomhole pressure (up to 52MPa), the change of gas compressibility is low, changing slightly. In conclusion, the pressure response time is sensitive to the wave velocity. Both the wave velocity and pressure response time are dominated by gas influx rate and pressure in the annulus, especially the gas influx rate. Within the range of high gas influx rate, the wave velocity decreases significantly.

CONCLUSIONS

A new method for predicting the position of gas influx in drilling operations based on PRP has been proposed. The mathematical model is solved by compiled code on VC++ (Version 2010) language. The main conclusions can be summarized as follows.

- In this paper, the pressure response time plate is calculated with full consideration of important factors which has influence on wave velocity. Experimental results show that the calculated pressure response time has good consistency with the experimental data.

- When gas influx occurs and migrates along the flow direction in the annulus from the bottomhole to wellhead, the wave velocity first slightly decreases and then sharply decreases. With the gas influx rate decreases or the BP increases, the wave velocity increases and pressure responses time decreases. Pressure response time is sensitive to the wave velocity. Both the wave velocity and pressure response time are dominated by gas influx rate and pressure in the annulus, especially the influx rate.

- The calculated PRP is unique at different gas influx position and gas influx rate. According to the corresponding principle, the gas influx position can be determined on the basis of the PRP with the known detected response time when gas influx occurs.

- Without the help of downhole tools, an accurate mathematical model to predict the position of gas influx based on PRP is of great importance and is feasible. The computer programming of mathematical model can be installed in the IPC to predict the position of gas influx in real time in drilling site. The new method provides accurate prediction of gas influx position in comparison with the field experiment. The prediction method is not only quickly and accurate, but it also saves drilling nonproductive time (NPT).

ACKNOWLEDGMENTS

Research work was cofinanced by the National Natural Science Foundation of China (no. 51274170) and Important National Science and Technology Specific Projects (2011ZX05022-005-005HZ). Without their support, this work would not have been possible.

REFERENCES

1.   Y. Bu, F. Li, Z. Wang, and J. Li, "Preliminary study on air injection in annuli to manage pressure during cementing," in Proceedings of the 2nd SREE Conference on Chemical Engineering (CCE '11), pp. 329–334, December 2011. · ·

2.   P. Vieira, F. Torres, R. A. Qamar, G. E. Marin, et al., "Down hole pressure uncertainties related to deep wells drilling are safely and precisely ascertained using automated MPD technology," in Proceedings of the North Africa Technical Conference and Exhibition, Society of Petroleum Engineers, 2012.

3.   S. Saeed, R. Lovorn, and K. Arne Knudsen, "Automated drilling systems for MPD C-the reality," inProceedings of the IADC/SPE Drilling Conference and Exhibition, 2012.

4.   J. A. Tarvin, I. Walton, P. Wand, and D. B. White, "Analysis of a gas kick taken in a deep well drilled with oil-based mud," in Proceedings of the SPE Annual Technical Conference and Exhibition, pp. 255–264, October 1991.

5.   W. Guo, F. Honghai, and L. Gang, "Design and calculation of a MPD model with constant bottom hole pressure," Petroleum Exploration and Development, vol. 38, no. 1, pp. 103–108, 2011. · ·

6.   J. Choe and H. Juvkam-Wold, "A modified two-phase well-control model and its computer applications as training and educational tool," SPE Computer Applications, vol. 9, no. 1, pp. 14–20, 1997.

7.   W. A. Bacon, "Consideration of Compressibility Effects for Applied-back-pressure Dynamic Well Control Response to a Gas Kick in Managed Pressure Drilling Operations," 2011.

8.   P. Vieira, F. Torres, R. Qamar, et al., "Down hole pressure uncertainties related to deep wells drilling are safely and precisely

ascertained using automated MPD technology," in Proceedings of the North Africa Technical Conference and Exhibition, 2012.

9. S. Wang, C. Qiang, and K. Bo, "Fluctuating pressure calculation during the progress of trip in managed pressure drilling," Advanced Materials Research, vol. 468–471, pp. 1736–1742, 2012. · ·

10. S. H. I. Yong quan, "Determining the Maximum Depth Method for Complex Formation Pneumatic DTH drilling with the Casing," Geology and Exploration, 4, 2009.

11. S. J. Chen and J. T. Aumann, "Numerical simulation of MWD pressure pulse transmission," inProceedings of the SPE Annual Technical Conference and Exhibition, 1985.

12. T. S. Collett, M. W. Lee, M. V. Zyrianova et al., "Gulf of mexico gas hydrate joint industry project leg II logging-while-drilling data acquisition and analysis," Marine and Petroleum Geology, vol. 34, no. 1, pp. 41–61, 2012. · ·

13. H. Chen, R. L. Brown, and J. P. Castagna, "AVO for one- and two-fracture set models," Geophysics, vol. 70, no. 2, pp. C1–C5, 2005. · ·

14. B. Joyce, D. Patterson, J. V. Leggett, et al., "Introduction of a new omni-directional acoustic system for improved real-time LWD sonic logging-tool design and field test results," in Proceedings of the SPWLA 42nd Annual Logging Symposium, 2001.

15. P. Radzinski and D. B. LWD, "MWD tools overcoming HP/HT demands," American Oil and Gas Reporter, vol. 47, no. 7, pp. 71–74, 2004.

16. C. R. Chia and B. C. De Lima, "MWD survey accuracy improvements using multistation analysis," inProceedings of the IADC/SPE Asia Pacific Drilling Technology Conference and Exhibition, pp. 143–150, September 2004.

17. H. Wang, S. B. M. Beck, G. H. Priestman, and R. F. Boucher, "Fluidic pressure pulse transmitting flowmeter," Chemical Engineering Research and Design, vol. 75, no. 4, pp. 381–391, 1997.

18. A. R. Young, MWD data transmission: U.S. Patent 8, no. 509, pp. 2–21, 2012.

19. M. Geraud, I. Butt, N. Evans, et al., "New generation PDC bits set new benchmarks in carbonate drilling, resulting in significant

performance improvements and cost savings for the operator," in Proceedings of the SPE Middle East Unconventional Gas Conference & Exhibition, 2013.

20. V. D. Better, "Well control through safe drilling margin identification, influx analysis and direct measurement method for deepwater," in Proceedings of the Offshore Technology Conference, vol. 2013, 2013.

21. J.-W. Park, D. A. Drew, and R. T. Lahey Jr., "The analysis of void wave propagation in adiabatic monodispersed bubbly two-phase flows using an ensemble-averaged two-fluid model," International Journal of Multiphase Flow, vol. 24, no. 7, pp. 1205–1244, 1999.
· ·

22. J. Xu and T. Chen, "Acoustic wave prediction in flowing steam-water two-phase mixture," International Journal of Heat and Mass Transfer, vol. 43, no. 7, pp. 1079–1088, 2000. · ·

23. F. Huang, M. Takahashi, and L. Guo, "Pressure wave propagation in air-water bubbly and slug flow,"Progress in Nuclear Energy, vol. 47, no. 1–4, pp. 648–655, 2005. ·

24. B. Bai, L. Guo, and X. Chen, "Pressure fluctuation for air-water two-phase flow," Journal of Hydrodynamics, vol. 18, no. 4, pp. 476–482, 2003.

25. I. Zubizarreta, "Pore pressure evolution, core damage and tripping out Schedulesf: a computational fluid dynamics approach," in Proceedings of the SPE/IADC Drilling Conference and Exhibition, pp. 5–7, Amsterdam, The Netherlands, March 2013.

26. C. H. Whitson, "Cyclic shut-in eliminates liquid-loading in gas wells," in Proceedings of the SPE/EAGE European Unconventional Resources Conference and Exhibition, pp. 20–22, Vienna, Austria, March 2012.

27. G. M. de Oliveira, A. Teixeira Franco, C. O. R. Negrao, A. Leibsonhn Martins, and R. A. Silva, "Modeling and validation of pressure propagation in drilling fluids pumped into a closed well," Journal of Petroleum Science and Engineering, vol. 103, pp. 61–71, 2012.

28. Y. Sato and H. Kanki, "Formulas for compression wave and oscillating flow in circular pipe," Applied Acoustics, vol. 69, no. 1, pp. 1–11, 2008. · ·

29. H. Li, Y. Meng, G. Li, et al., "Propagation of measurement-while-drilling mud pulse during high temperature deep well drilling operations," Mathematical Problems in Engineering, vol. 2013, Article ID 243670, 12 pages, 2013. ·

30. X. Wang and J. Zhang, "The research of pressure wave Pulsation in mud pulse transmitting," Journal of Chongqing University of Science and Technology, vol. 14, no. 2, pp. 55–58, 2012.

31. Y. Lin, X. Kong, Y. Qiu, et al., "Calculation analysis of pressure wave velocity in gas and drilling mud two-phase fluid in annulus during drilling operations," Mathematical Problems in Engineering, vol. 2013, Article ID 318912, 17 pages, 2013. ·

32. J. Orkiszewski, "Predicting two-phase pressure drops in vertical pipe," Journal of Petroleum Technology, vol. 19, no. 6, pp. 829–838, 1967.

33. X. Kong, Y. Lin, Y. Qiu, et al., "A new model for predicting dynamic surge pressure in gas and drilling mud two-phase flow during tripping operations," Mathematical Problems in Engineering, vol. 2014, Article ID 916798, 16 pages, 2014. ·

Chapter 2

Surface Modification by Friction Based Processes

R. M. Miranda[1], J. Gandra[2], and P. Vilaça[2]

[1]Mechanical and Industrial Engineering Department, Sciences and Technology Faculty, Nova University of Lisbon, Caparica, Portugal

[2]Mechanical Engineering Department, Lisbon Technical University, Av. Rovisco Pais, Lisboa, Portugal

INTRODUCTION

The increasing need to modify the surface's properties of full components, or in selected areas, in order to meet with design and functional requirements, has pushed the development of surface engineering which is largely recognised as a very important field for materials and mechanical engineers.

Surface engineering includes a wide range of processes, tailoring chemical and structural properties in a thin surface layer of the substrate, by modifying the existing surface to a depth of 0.001 to 1.0 mm such as: ion implantation, sputtering to weld hardfacings and other cladding processes, producing typically 1 - 20 mm thick coatings, usually for wear and corrosion resistance and repairing damaged parts. Other deposition processes, such as laser alloying or cladding, thermal spraying, cold spraying, liquid deposition methods, anodising, chemical vapour deposition (CVD), and physical vapour deposition (PVD), are also extensively used in surface engineering. Hardening by melting and rapid solidification and surface mechanical deformation allow to change the properties without modifying its composition [1].

Friction based processes comprise two manufacturing technologies and these are: Friction Surfacing (FS) and Friction Stir Processing (FSP). The former was developed in the 40´s [2] and was abandoned, at that time, due to the increasing developments observed in competing technologies as thermal spraying, laser and plasma. Specially laser surface technology has largely developed in the following years in hardening, alloying and cladding applications and is now well established in industry. However, FS as a solid state processing technology, was brought back for thermal sensitive materials due to its possibility to transfer material from a consumable rod onto a substrate producing a coating with a good bonding and limited dilution.

The patented concept of Friction Stir Welding in the 90´s [5] opened a new field for joining metals, specially light alloys and friction stir processing emerged around this concept.

FSP uses the same basic principles as friction stir welding for superficial or in-volume processing of metallic materials. Applications are found in localized modification and microstructure control in thin surface layers of processed metallic components for specific property enhancement. It has proven to be an effective treatment to achieve major microstructural refinement, densification and homogenisation of the processed zone, as well as, to eliminate defects from casting and forging [6-8]. Processed surfaces have enhanced mechanical properties, such as hardness, tensile strength, fatigue, corrosion and wear resistance. A uniform equiaxial fine grain structure is obtained improving superplastic behaviour. FSP has also been successfully investigated for metal matrix composite manufacturing (MMCs) and functional graded materials (FGMs) opening new possibilities to chemically modify the surfaces [9].

However, FSP has some disadvantages, the major of which is tool degradation and cost, which limits its wider use to high added value applications. Therefore, friction surfacing (FS) emerged again.

This chapter will focus on the mechanisms involved in both FSP and FS and their operating parameters, highlighting existing and envisaged applications in surface engineering, based on the knowledge acquired from ongoing research at the author's institutions.

FRICTION STIR PROCESSING

Fundamentals

Friction Stir Processing (FSP) is based on the same principles as friction stir welding (FSW) and represents an important breakthrough in the field of solid state materials processing.

FSP is used for localized modification and microstructural control of surface layers of processed metallic components for specific property enhancement [6]. It is an effective technology for microstructure refinement, densification and homogenisation, as well as for defect removal of cast and forged components as surface cracks and pores. Processed surfaces have shown an improvement of mechanical properties, such as hardness and tensile strength, better fatigue, corrosion and wear resistance. On the other hand, fine microstructures with equiaxed recrystallized grains improve superplastic behaviour of materials processing and this was verified for aluminium alloys [7]. More recently the introduction of powders preplaced on the surface or in machined grooves allowed the modification of the surfaces, producing coatings with characteristics different from the bulk material, or even functionally graded materials to be discussed later in this chapter. The process has still limited industrial applications but is promising due to its low energy consumption and the wide variety of coating / substrate material combinations allowed by the solid state process.

A non-consumable rotating tool consisting of a pin and a shoulder plunges into the workpiece surface. The tool rotation plastically deforms the adjacent material and generates frictional heat both internally, at an atomic level, and between the material surface and the shoulder. Localized heat is produced by dissipation of the internal deformation energy and interfacial friction between the rotating tool and the workpiece. The local temperature of the substrate rises to the range where it has a viscoplastic behaviour beneficial for thermo-mechanical processing. When the proper thermo-mechanical conditions, necessary for material consolidation are achieved, the tool is displaced in a translation movement. As the rotating tool travels along the workpiece, the substrate material flows, confined by the rigid tool and the adjacent cold material, in a closed matrix like forging manufacturing process. The material under the tool is stirred and forged by the pressure exerted by the axial force applied during processing as depicted in Figure 1.

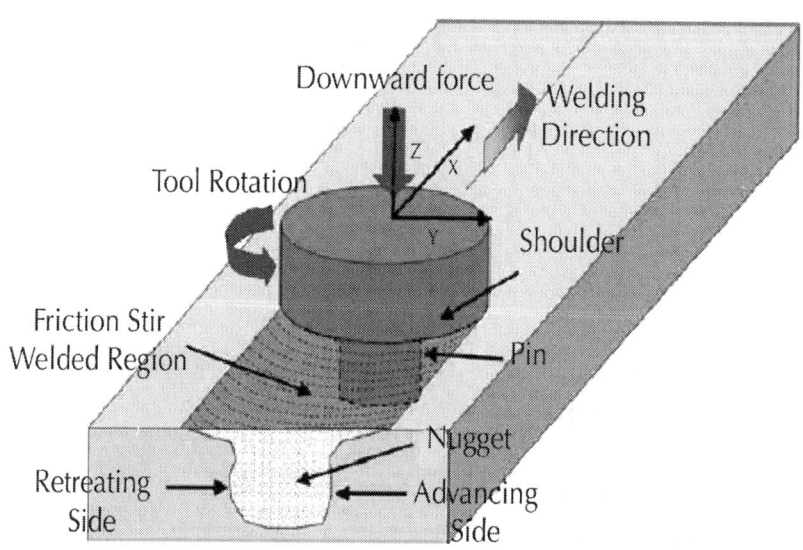

Figure 1: Schematics of friction stir processing.

The material structure is refined by a dynamic recrystallization process triggered by the severe plastic deformation and the localised generated heat. Homogenization of the structure is also observed along with a defect free modified layer of micrometric or nanometric grain structure.

FSP is considered an environmentally friend technology due to its energy efficiency and absence of gases or fumes produced. Table 1 summarizes the major benefits of FSP considering technical, metallurgical, energy and environment aspects.

Table 1: Major benefits of friction stir processing

Technical	Processed depth controlled by the pin length
	One-step technique
	No surface cleaning required
	Good dimensional stability since it is performed under solid state
	Good repeatability
	Facility of automation
Metallurgical	Solid state process
	Minimal distortion of parts
	No chemical effects
	Grain refining and homogenization
	Excellent metallurgical properties
	No cracking
	Possibility to treat thermal sensitive materials
Energy	Low energy consumption since heat is generated by friction and plastic deformation
	Energy efficiency competing with fusion based processes as laser
Environmental	No fumes produced
	Reduced noise
	No solvents required for surface degreasing and cleaning

Analysing the cross section of a friction stir processed surface, three distinct zones can be identified and these are: the nugget or the stirred zone (SZ), the thermomechanically affected zone (TMAZ) and the heat affected zone (HAZ) as shown in Fig. 2.

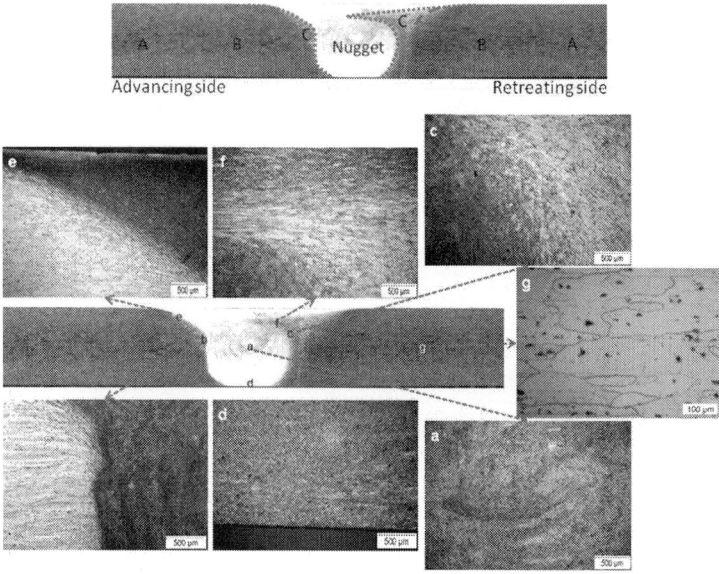

Figure 2: A typical macrograph showing various microstructural zones in FSW of AA2024-T351.

The nugget, just below the pin and confined by the shoulder width, is the area of interaction where severe plastic deformation occurs. The raise in local temperature due to internal friction and the generated friction between the shoulder and the surface along with the high strain, promotes a dynamically recrystallized zone, resulting in the generation of fine homogeneous equiaxial grains in the stirred zone and precipitate dissolution. Though this is a solid state process, the maximum temperature can be of about 80% of the fusion temperature. Ultrafine-grained microstructures with an average grain size of 100-300 nm in a Mg-Al-Zn alloy were observed in a single pass under cooling [8]. These micro and nano structures are responsible for increases in hardness and wear behaviour reported by several researchers studying different types of alloys under different processing conditions.

The thermo-mechanically affected zone (TMAZ) is immediately adjacent to the previous and, in this zone, the deformation and generated heat are insufficient to generate new grain formation, thus, deformed elongated grains are observed with second phases dispersed in the grain boundaries. Though new grain nucleation may be observed, microstructure remains elongated and deformed. The hardness is higher than in the heat-affected zone due to the high dislocations density and sub-boundaries caused by plastic deformation.

In the heat affected zone (HAZ) no plastic deformation is experienced but heat dissipated from the stirred zone into the bulk material can induce phase transformations, depending on the alloys being processed, as precipitate coarsening, localized aging or annealing phenomena.

The non symmetrical character of the process is also evident (Fig. 2). The advancing side is usually referred to the one where the rotating and travel movements have the same direction, while in the retreating side these have opposite directions. On the advancing side, the recrystallized zone is extended and the nugget presents a sharp appearance. The relative velocity between the tool and the base material is higher due to the combination of tool rotation and translation movement. As such, plastic deformation is more intense, thus, the increase in the degree of deformation during FSP, results in a reduction of recrystallized grain size, extending the fine-grain nugget region to the advancing side. Hardness can be higher than in the thermo-mechanically affected zone, but typically lower than in the base material, whenever it is a heat treatable alloy hardenable by aging. The Hall Petch equation establishes a relation between grain size and yield strength and states that these vary in opposite senses [10]. So, in the nugget yield strength is seen to be much higher than in the base material and this is a major result from this process.

Processing Parameters

Operating or processing parameters determine the amount of plastic deformation, generated heat and material flow around the non-consumable tool.

The tool geometry is of major relevance as far as material flow is concerned. Two main elements constitute the tool and these are the pin and the shoulder. Geometrical features such as pin height and shape, shoulder surface pattern and diameter, have a major influence on material flow, heat generation and material transport volume, determining the final microstructure and properties of a processed surface. Several tools have been designed and patented for both FSW and FSP.

The pin (Fig. 3) can be cylindrical or conical, flat faced, threaded or fluted to increase the interface between the probe and the plasticized material, thus intensifying plastic deformation, heat generation and material mixing. The pin length determines the depth of the processed layer. However, since FSP usually aims to produce a thin fine-grained layer across a larger surface area, pinless tools with larger shoulder diameters can also be used.

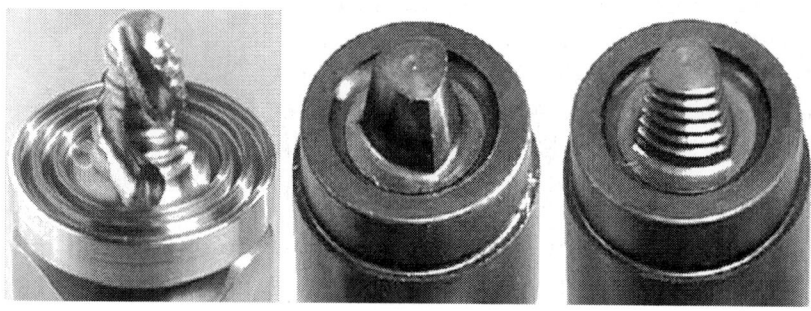

Figure 3: Example of tool geometries.

Shoulder profiles aim to improve friction with the material surface generating the most part of frictional heat involved in the process. Shoulders can be concave, flat or convex, with grooves, ridges, scrolls or concentric circles as depicted in Figure 4.

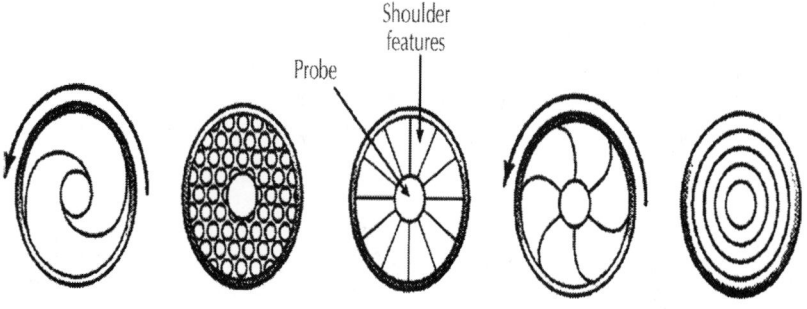

Figure 4: Examples of shoulder geometries.

The major processing parameters are the tool rotation and traverse speeds, the axial force and the tilt angle:

- Tool rotation and traverse speeds

These parameters, individually or in combination, affect the plastic deformation imposed onto the material and, thus, the generated heat. An empirically accepted concept divides processing into two main classifications: "cold" and "hot". Cold processing is the one where the ratio between rotating and traverse speed is below 3 rpm/mm and hot processing when this ratio is above 6. Though there is no scientific basis for this border line, it is, however, noticeable that increasing this ratio, the SZ is larger and a very fine structure is observed, while under "cold" conditions the SZ is not well defined since the heat generated is insufficient to promote grain recrystallization. So, increasing the tool rotation speed, plastic deformation is more intense and so is generated heat enabling more material mixing. Therefore, it is possible to achieve a smaller grain size of equiaxial homogeneous grains with precipitate dissolution.

Transverse speed mostly affects the exposure time to frictional heat and material viscosity. Low traverse speeds result in larger exposure times at higher process temperatures.

- Tool axial force

This parameter affects friction between the shoulder and the substrate surface generating and promoting material consolidation. High axial force causes excessive heat and forging pressure, obtaining grain growth and coarsening, while low axial forces lead to poor material consolidation, due to insufficient forging pressure and friction heating. Excessive force may also result in shear lips or flashes with excessive height of the beads on both the advancing and retreating sides, causing metal thinning at the processed area and poor yield and tensile properties. So, surface finishing is much controlled by the axial or forging force.

- Tilt angle

The tilt angle is the angle between the tool axis and the workpiece surface. The setting of a suitable tilting towards the traverse direction assures that the tool moves the material more efficiently from the front to the back of the pin and improves surface finishing.

The effect of the different process parameters has been widely documented by several authors and they are all unanimous that plastic deformation and consequent heat generation are essential to establish the viscoplastic conditions necessary for the material flow and to achieve good consolidation. Thus, a tilt angle of 2-4° is usually used in practice.

Insufficient heating, caused by poor stirring (low tool rotational rates), a high transverse speed or insufficient axial force, results in improper material consolidation with consequent low strength and ductility. Raising heat will cause grain size to decrease to a nanometric scale improving material properties. However, a very significant increase in tool rotation rate, axial force or a very low transverse speed may result in high non desired temperature, slow cooling rate or excessive release of stirred material with property degradation.

Multiple Passes

In order to process large areas in full extent, multiple-passes are used. These can be run separately or overlapped. An overlap ratio (OR) was defined to characterize the overlap between passes and defined by equation 1 [7].

$$OR = 1 - [l dpin]$$

(1)

Where l is distance between centres of each pass and d_{pin} is the maximum diameter of the pin. From this equation, fully overlapped passes have an OR=1 and OR decreases, when increasing the distance between passes. For an OR<0 no overlap of the nuggets exists.

There are two types of material modification by Friction Stir Processing, the in-volume FSP (VFSP) consisting on the modification of the full thickness of the processed materials and the surface FSP (SFSP) which consists in the surface modification up to depth of about 2 mm.

Figure 5 depicts the effect of OR in two Al alloys, a heat treatable (AA7022-T6) and a non heat treatable one (AA5083-O) with different number of passes and overlap ratios.

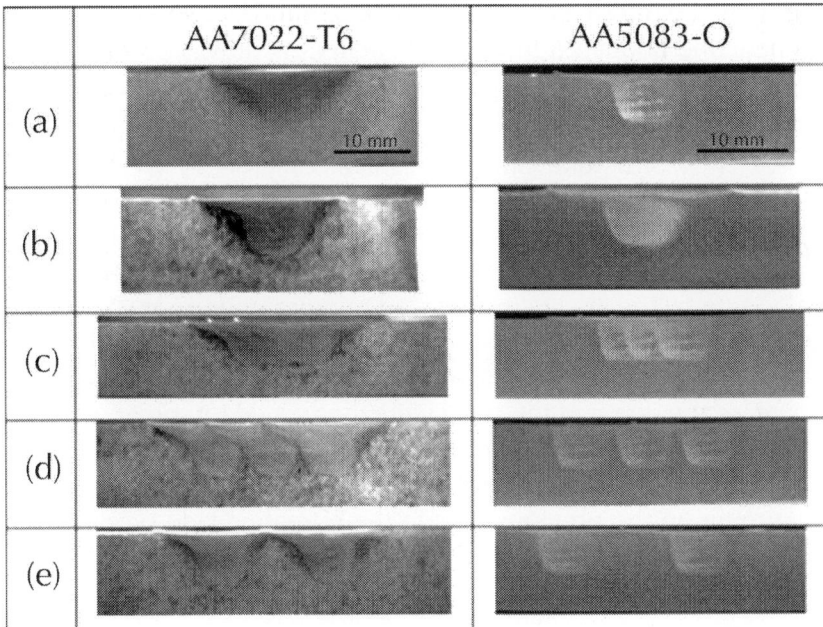

Figure 5: Cross sections of the samples processes with different treatments a) one pass with OR=1; b) four passes with OR=1; c) three passes with OR=1/2; d) three passes with OR=0 and e) two passes with OR=-1 [7].

In this study [7] the authors showed that AA5083-O alloy needed at least three passes in the same location to produce a homogeneous processed area, while the AA7022-T6 alloy only needed one pass, since this is a heat treatable alloy. Grain size reduced from 160 μm (AA7022-T6) and 106 μm (AA5083-O) to an average grain size of about 7.1 and 5.9 μm, respectively. The highest hardness value was located in the nugget due to a significantly decrease in the grain size. This results that in AA7022-T6 alloy the hardness is lower in the nugget than in the base material because it is a heat treatable aluminium alloy and in the AA5083-O alloys the hardness in the nugget is higher than in the base material which is a typical behaviour of non-heat treatable alloys. A significant increase in the formability of the materials was observed due to the increase of the materials ductility resulting from the refinement of the grain size, increasing the maximum bending angle in four times for the SFSP treatment and twelve times for the VFSP treatment in the AA7022-T6 samples. In AA5083-O samples an increase in the maximum bending angle around 1.5 times for the SFSP treatment and about 2.5 times for VFSP treatment was observed.

The overlapping direction in multipass Friction Stir Processing (FSP) was also seen to have a major influence on the surface geometrical features [11]. Structural and mechanical differences were observed in a AA5083-H111 alloy when overlapping by the advancing side (AS) direction or by the retreating side (RS) one. Overlapping by the retreating side was found to generate smoother surfaces, while overlapping by the advancing side led to more uniform thickness layer (Fig. 6). This result is quite relevant from a practical point of view since when the aim of processing large areas in multiple passes procedure is to increase the depth of the processed zone, overlapping of successive passes should be performed by the advancing side of the previous pass. If surface finishing is to be maximised to prevent finishing operations, overlapping on the previous pass in the retreating side produces very low rough surfaces.

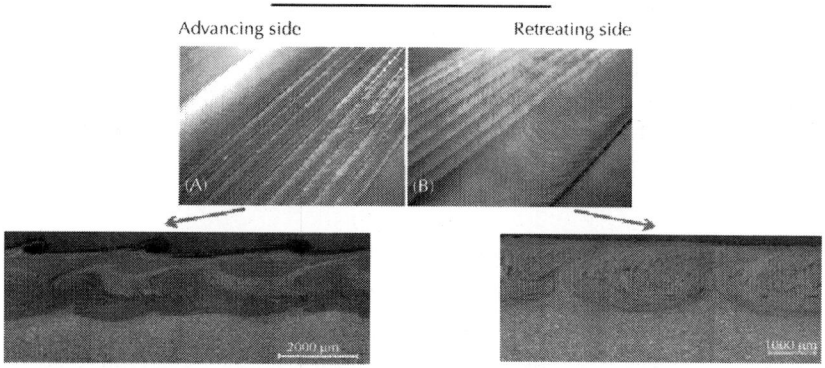

Figure 6: Macro and micrographs of cross sections in friction stir processed surfaces when overlapping by the advancing and by the retreating sides [11].

Hardness within the processed layer increased by 8.5 % and was seen to be approximately constant between passes. The mechanical resistance and toughness under bending were improved by 18 % and 19 %, respectively.

Bending test curves are presented in Figure 7. The processed surfaces were tested under tensile and compression loads. Different behaviours were observed for each bending specimen. Surface modification by multi-pass FSP resulted in an increase of the maximum load supported for all samples and up to a maximum of 18 % for the compression solicitation of the sample produced when overlapping by the RS (Fig. 8). FSP produced a thin layer of a fine equiaxial recrystallized grain structure and homogeneous precipitation dispersion, enhancing material strength.

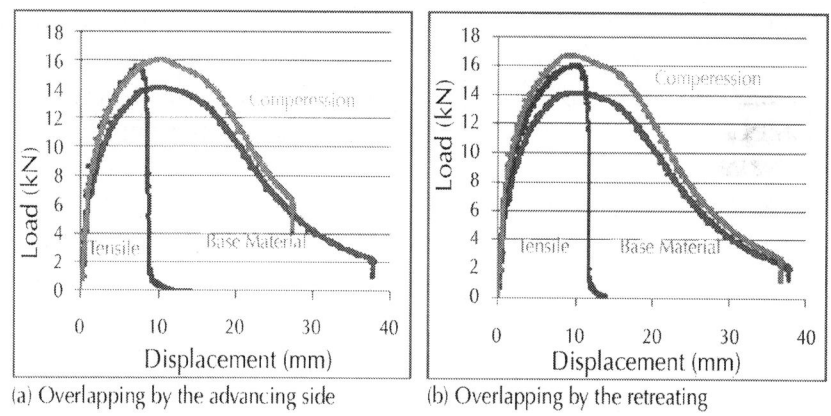

(a) Overlapping by the advancing side   (b) Overlapping by the retreating

Figure 7: Load vs. displacement plot of the bending tests of the FSP samples produced when overlapping by (a) AS and (b) RS [11].

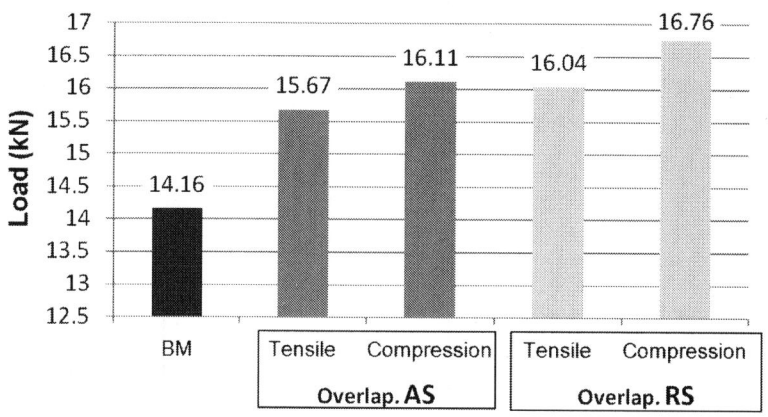

Figure 8: Maximum load attained by FSP samples under different test conditions relatively to the base material [11].

Applications and Performance

Friction stir processing can be used to locally refine microstructures and eliminate casting defects in selected locations, where property improvements can enhance component performance and service lifetime. For instances, aluminium castings contain porosities, segregated phases and inhomogeneous microstructures which contribute to property degradation. Microstructural casting defects, such as: coarse precipitates and porosities increase the possibility of rupture due to the intragranular nucleation of micro-cracks during material deformation. Precipitates are less capable of plastic deformation than the matrix, so cavity nucleation is very frequent, whether caused by a disconnection from the matrix or the rupture of precipitates.

Friction stir processing allows the breakage of large precipitates and their dispersion in a homogeneous matrix, increasing the material capability to withstand deformation, since it results in a higher level of crack closure. Additionally, mechanical properties such as ductility, fatigue strength and formability, are improved.

On the other hand, a large number of small precipitates increases the material resistance to deformation and hence its strength, as they act as barriers or anchorage points to dislocations movements. A uniform equiaxial fine grain structure is also essential to enhance material superplastic behaviour. Friction stir processing generates fine microstructure and equiaxed recrystallized grains which leads either to an increase in strain rate or a decrease in the temperature at which superplasticity is achieved.

In the FSP of an aluminium cast alloy ADC12, Nakata et al. [12] applied multiple-passes to increase tensile strength to about 1.7 times that of the base material. The hardness profile of processed layer was uniform and about 20 HV higher than that of the cast material. The observed increase in tensile strength was attributed to the elimination of the casting defects such as porosities, an homogeneous redistribution of fine Si particles and a significant grain refinement to 2–3 µm. Santella et al. [13] investigated the use of friction stir processing to homogenise hardness distributions in A319 and A356 cast aluminium alloys. Hardness and tensile strength were increased relatively to the cast base material.

Similar results were also reported in the friction stir processing of magnesium based alloys. A.H. Feng and Z.Y. Ma et al. [14] combined FSP with subsequent aging to enhance mechanical properties of Mg-Al-Zn castings.

Chang et al. [8] obtained a significant improvement of mechanical properties as the mean hardness measured at the ultrafine-grained zone reached approximately 120HV (twice the base material hardness).

Several investigations have been conducted to study the enhancement of superplasticity behaviour in friction stir processed alloys. In the FSP of Al-8.9Zn-2.6Mg-0.09Sc, Charit and Mishra et al. [15] reported a maximum superplasticity of 1165% at a strain rate of 3×10^{-2} s^{-1} and 310 °C with a grain size of 0.68 µm. More recently, F.C. Liu [16] reported a fine-grain microstructure of 2.6 µm sized grains by applying FSP to extruded samples of an Al-Mg-Sc alloy, achieving a maximum elongation of 2150% at a high strain rate of 1×10^{-1} and a temperature of 450 °C.

A ultrafine-grained FSP Al-Mg-Sc alloy was also reported [17] with a grain size of 0.7 µm exhibited high strain rate superplasticity, for a low temperature range of 200 to 300 °C with a single pass. For a strain rate of 3×10^{-2} s^{-1} at a temperature of 300 °C, a maximum ductility of 620% was achieved. However, for a temperature of 350 °C, abnormal grain growth was observed, as grain size increased and the samples no longer presented superplasticity, thus confirming that grain size is essential for the existence of a superplastic behaviour. García-Bernal et al. [18] conducted a study to evaluate the high strain rate superplasticity behaviour during the high-temperature deformation of a continuous cast Al-Mg alloy, having reported that the generation of a fine grain structure and the breaking of cast structure led to a significant improvement in its ductility up to 800% at 530 °C and a strain rate of 3×10-2 s-1.

The fine-grained microstructure generated by FSP can also prevent fatigue crack initiation and propagation due to the barrier effect of grain boundaries. For example, Jana et al. [19] friction stir processed a cast Al-7Si-0.6Mg alloy, widely used for its good castability, mechanical properties and corrosion resistance, but characterized by poor fatigue properties. The authors succeeded to improve fatigue resistance by a factor of 15 at a stress ratio of R=$\sigma_{min}/\sigma_{max}$ = 0 due to a significant enhancement of ductility and a homogeneous redistribution of refined Si particles.

Intense plastic deformation and material mixing featured in the FSP of A356 aluminium casting also resulted in the significant breakage of primary aluminium dendrites and coarse Si particles, creating a homogenous distribution of Si particles in the aluminium matrix and eliminating casting porosity [7]. This led to a significant improvement of ductility and fatigue strength in 80%, proving that FSP can be used as a tool to locally modify the microstructures in regions experimenting high fatigue loading.

Friction surfacing of AA6082-T6 over AA2024-T3 evidenced a significant improvement of wear performance in about 25 %, compared to the consumable rod in as-received condition (Table 1). This enhancement in wear behaviour is also due to a finer equixial grain microstructure within the coating, compared to the rod anisotropic microstructure which is more prone to delamination under wear loads.

AA2024-T3 substrate plates exhibited the best tribological properties, presenting the lowest weight loss, frictional force and friction coefficient. This is most likely due to both its higher surface hardness and its lower ductility, which make this material less prone to suffer plastic deformation under abrasive wear, in comparison with the AA6082 coating and the rod in as-received condition. Due to the fine grain structure observed, the coatings present high frictional force and coefficient (10.9 N and 0.56, respectively).

Table 2: Weight loss due to wear (average values)

Material	Weight lost [mg]	Volume lost [mm3]	Volume rate [10-2 mm 3/m]	Wear rate [mg/m]	First stage Run-in wear		Second stage Steady state wear	
					Frictional force [N]	Friction coefficient	Frictional force [N]	Friction coefficient
Substrate	12.6 ± 3	4.54	1.51	0.042	4.9 ± 0.97	0.25 ± 0.05	7.5 ± 0.33	0.38 ± 0.017
ARCR	30.2 ± 5	11.19	3.73	0.101	-	-	7.1 ± 1.14	0.36 ± 0.059
Coatings	23.2 ± 3	8.59	2.86	0.077	-	-	10.9 ± 0.58	0.56 ± 0.029

Surface Composites

FSP has been also investigated to produce layers of hard materials on soft substrates, as aluminium based alloys. Most of the published work is focused on the effect of processing parameters on surface characteristics and techniques to evaluate the performance of modified surfaces. Nevertheless, the reinforcing particles deposition method is relevant in terms of structural and chemical homogeneity and depth of the modified layer which influence the final surface performance. Different methods for depositing reinforced particles have been reported. A main reinforcing method consists of mixing reinforcing particles or powders with a volatile solvent such as methanol or a lacquer, in order to form a thin reinforcement layer, preventing reinforcing powders to escape. Another method consists of machining grooves in the substrate, pack these with reinforcing particles and process the zone with a non consumable FSP tool in a single pass or in multiple passes.

An enormous diversity of materials is used for surface reinforcements, the majority being hard ceramic particles as SiC, Al_2O_3 and AlN to improve surface properties as hardness, superplasticity, formability, corrosion and wear resistances.

The paricle size is relevant since small particles lead to higher concentration along bead surface and to smooth fraction gradients both in depth and along the direction parallel to the surface, while the thickness of the reinforced layer decreases with increasing particle size and is, typically, below 100 micron [20].

More recently, nanostructured layers have been produced and less common reinforcements were studied successfully. Two examples are: the incorporation of multi-walled carbon nanotubes (MWCNT) into a number of metallic materials as reinforcing fibres is a topic of recent interest due to the unique mechanical and physical properties of this material, namely very high tensile strengths [21]. FSP was tested to produce a composite of an aluminium alloy with MWCNT. Nanotubes were embedded in the stirred zone and the multi walled was retained. With tool rotational speeds of 1500 and 2500 rpm the distribution of nanotubes increased. Aiming at weight reduction of vehicles, FSP MWCNT/AZ31 surface composite were produced by Morisada et al. [22] and succeeded to disperse MWCNT into a AZ31 matrix. The microhardness increased to values of about 74 HV and the addition of MWCNT was seen to further promote grain refinement by FSP.

Another example is the incorporation of Nitinol (NiTi) that is a shape memory alloy with superelastic behavior and good biocompatibility. These alloys are widely used in orthodontics, but also in sensors and actuators. The possibility of incorporating wires, ribbons or powders into metallic matrixes opens up new applications for shape memory alloys. Studies report on the use of NiTi wires, but few have been made in the dispersion of NiTi powders in a metal matrix. Dixit et al. [23] produced a NiTi reinforced AA1100 composite using FSP and the particles were uniformly distributed. Good bonding with the matrix was achieved and no interfacial products were formed. The authors suggest that under adequate processing, the shape memory effect of NiTi particles can be used to induce residual stress in the parent matrix, of either compressive or tensile type. This study showed that samples had enhanced mechanical properties such as: Young modulus and micro hardness. A more recent work showed the possibility to introduce 1x2 mm ribbons of NiTi in AA1050 alloy by FSP showing a good vibration and damping capacity of the composite [24].

Shafei-Zarghani et al. [25] used multiple-pass FSP to produce a superficial layer of uniformly distributed nano-sized Al2O3 particles into an AA6082 substrate. Hardness was increased three times over that of the base material. Wear testing revealed a significant resistance improvement. Researchers also found that the increase of the number of passes leads to more uniform alumina particle distributions with a significant increase of surface hardness. The nano-size Al2O3 powder was inserted inside a groove with 4 mm depth and 1 mm width, which was closed by a tool with a shoulder and no pin.

FRICTION SURFACING

Principles and Process Parameters

Friction surfacing (FS) was first patented in the 40´s and is now well established as a solid state technology to produce metallic coatings. While FSP modifies the microstructure of a surface by simply deforming, recrystallize and homogenise the grain structure, FS modifies its chemistry. In friction surfacing a consumable rod under rotation is pressed under an axial load against the surface as depicted in Fig. 9.

Heat generated in the initial friction contact promotes viscoplastic deformation at the tip of the rod. As the consumable travels along the substrate, the viscoplastic material at the vicinity of the rubbing interface flows into flash or is transferred over onto the substrate surface, while pressure and heat conditions triggers an inter diffusion process that soundly bonds the deposit. As the material undergoes a thermo-mechanical process, a fine grain microstructure is also produced by dynamic recrystallization.

Figure 9: Metallic coating of steel substrate by FS.

Gandra et al [20] proposed a model for the global thermal and mechanical processes involved during friction surfacing based on the metallurgical transformations observed when depositing mild steel over mild steel and is shown in Fig.10.

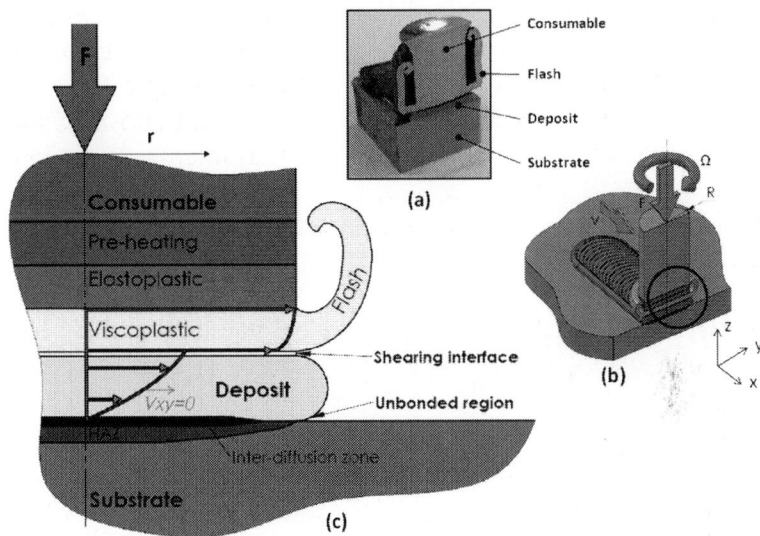

Figure 10: Thermo-mechanics of friction surfacing. (a) Sectioned consumable, (b) Process parameters and (c) Thermo-mechanical transformations and speed profile Nomenclature: F – Forging force; Ω – rotation speed; v -travel speed; Vxy – rod tangential speed in-plan xy given by composition of rotation and travel movements [20].

The speed difference between the viscoplastic material, which is rotating along with the rod at v_{xy}, and the material effectively joined to the substrate ($v_{xy} = 0$), causes the deposit to detach from the consumable. This viscous shearing friction between the deposit and the consumable is the most significant heat source in the process.

Since the deposited material at the lower end is pressed without lateral confinement, it flows outside the consumable diameter, resulting into a revolving flash attached to the tip of the consumable rod and side unbounded regions adjacent to the deposit. Flash and unbonded regions play an important role as boundary conditions of temperature and pressure for the joining process.

Fig 11 shows typical material combinations tested using FS with successful results.

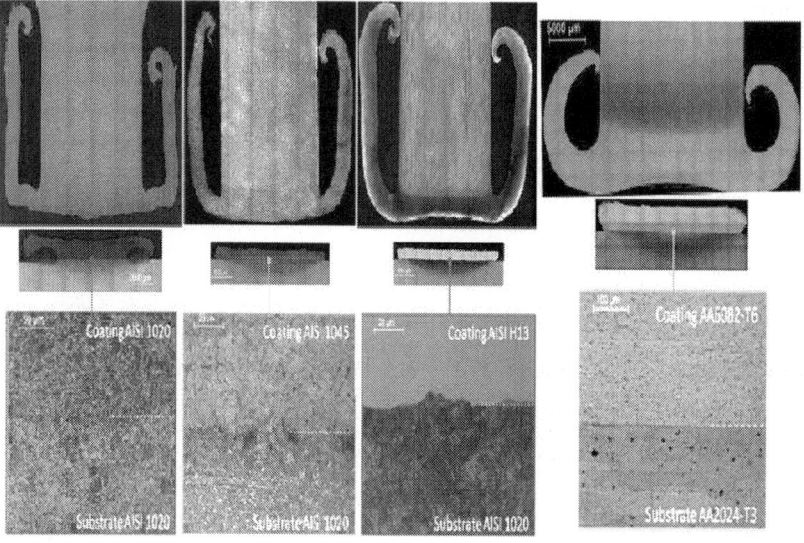

Figure 11: Different coatings/substrates combinations.

The process allows the deposition of various dissimilar material combinations as the deposition of stainless steel, tool steel, copper or Inconel on mild steel substrates, as well as, stainless steel, mild steel and inconel consumables on aluminium substrates.

The influence of processing parameters on the deposit characteristics and bonding strength has been studied [26,27] aiming to correlate the resulting coating geometrical characteristics (thickness and bonded width) and mechanical performance with forging force, spindle and travel speeds. The increase of forging force improves the bond strength and reduces the coating thickness. The undercut region decreased when the forging force increased and the travel speed decreased. Higher ratios between the consumable rod feeding rate and the travel speed resulted in superior bonding quality. The applied load on the consumable rod was found to be essential to improve joining efficiency and to increase the deposition rate. Higher rotation or travel speeds were detrimental for the joining efficiency. Tilting the consumable rod along the travel direction proved to improve the joining efficiency up to 5 %. The material loss in flashes represented about 40 to 60 % of the total rod consumed, while unbonded regions were reduced to 8 % of the effective coating section in mild steel deposition. Friction surfacing was seen to require mechanical work between 2.5 and 5 kJ/g of deposited coating with deposition rates of 0.5 to 1.6 g/s, that is, deposition rates are higher than for laser cladding or plasma arc welding and the specific energy consumption lower than for other cladding processes.

In the friction surfacing of low carbon steel with tool steel H13 consumable rods, Rafi et al. [28] concluded that the coating width was strongly influenced by the rotation speed, while thickness was mostly determined by the travel speed.

This field of exploitation of producing aluminium coatings on aluminium based alloys is very promising. It was seen that friction surfacing enables intermediate mass deposition rates and higher energy efficiency in comparison with several mainstream laser and arc welding cladding processes. The required mechanical work varied between 2.5 and 5 kJ/g of deposited coating with deposition rates of 0.5 to 1.6 g/s. The forging force enhances joining quality while contributing to a higher overall coating efficiency. Faster travel and rotation speeds improved deposition rates and coating hardness, while decreasing energy consumption per unit of mass. Surface hardness increased up to 115 % compared to consumable rod. By adjusting a proper tilt angle, specific energy consumption drops, while slightly improving deposition rate and joining efficiency.

SUMMARY AND FUTURE TRENDS

Friction based processes comprise Friction Stir Processing (FSP) and Friction Surfacing (FS).

Friction stir processing is mostly used to locally eliminate casting defects and refine microstructures in selected locations, for property improvements and component performance enhancement. Aluminium and steel castings are amongst the most common components improved by this technology aiming at eliminating porosities, destroy solidification structures with inhomogeneous segregated phases, refine grain structures improving n-service performance.

The recent advances in adding reinforcing particles to manufacture surface alloys and metal matrix composites is a breakthrough in this technology opening new possibilities to manufacture composites nanostructured with tremendous properties.

Friction surfacing has been used in the production of long-life industrial blades, wear resistant components, anti-corrosion coatings and in the rehabilitation of worn or damaged parts such as, turbine blade tips and agricultural machinery. Other applications feature the hardfacing of valve seats with stellite and tools such as punches and drills.

Since the deposits result from severe viscoplastic deformation, friction surfacing presents some advantages over other coating technologies based on fusion welding or heat-spraying processes, that produce coarse microstructures and lead to intermetallics formation, thereby deteriorating the mechanical strength of the coatings. However, friction surfacing currently struggles with several technical and productivity issues which contribute to a limited range of engineering applications.

REFERENCES

1.   Bunshah RF. Handbook of Deposition Technologies for Films and Coatings: Science, Technology and Applications. 2nd Edition. Noyes Publications; 1994.

2.   Klopstock H, Neelands AR. Patent specification, An improved method of joining or welding metals; Ref. 572789; 1941.

3.   Nicholas ED. Friction surfacing, ASM handbook: Vol 6. ASM International; 1993; p. 321–323.

4.   Bedford GM, Vitanov VI, Voutchkov II. On the thermo-mechanical events during friction surfacing of high speed steels. Surface and Coatings Tecnhology 2001; 141 (1) 34-39.

5.   Thomas W. Friction Stir But Welding, International Patent Application N° PCT/GB92/02203 and GB Patent Application N° 9125978.8, US Patent N°5460317.

6.   Mishra RS, Ma ZY. Friction stir welding and processing. Materials Science & Engineering R 2005; 50 (1-2) 1-78.

7.   Nascimento F, Santos T, Vilaça P, Miranda RM, Quintino L. Microstructural modification and ductility enhancement of surfaces modified by FSP in aluminium alloys. Materials Science and Engineering A 2009; 506 (1-2) 16-22.

8.   Chang ADu, XH, Huang JC, Achieving ultrafine grain size in Mg-Al-Zn alloy by friction stir processing. Scripta Materiallia 2007; 57 (3) 209-212.

9.   Gandra J, Miranda R, Vilaca P, Velhinho A, Pamies-Teixeira J. Functionally graded materials produced by friction stir processing. Journal of Materials Processing Technology 2011; 211 (11) 1659-1668.

10.   Porter DA, Easterling KE, Phase Transformations in Metals and Alloys 1992, CRC Press.

11.   Gandra J, Miranda RM, Vilaça P. Effect of overlapping direction in multipass friction stir processing. Materials Science and Engineering A 2011; 528 (16-17) 5592–5599.

12.   Nakata K, Kim YG, Fuji H, Tusumura T, Komazaki T. Improvement of mechanical properties of aluminium die casting alloy by multi-pass friction strir processing. Materials Science and Engineering A 2006; 437 (2) 274-280.

13.   Santella ML, Engstrom T, Storjohann D, Pan TY. Effects of friction stir processing on mechanical properties of the cast aluminum alloys A319 and A356. Scripta Materialia 2005; 53 (2) 201-206;

14.   Feng AH, Ma ZY. Enhanced mechanical properties of Mg-Al-Zn cast alloy via friction stir processing. Scripta Materiallia 2007; 56 (5) 397-400.

15.   Charit I, Mishra RS, Low temperature superplasticity in a friction-stir-processed ultrafine grained Al-Zn-Mg-Sc alloy. Acta Materialia 2005; 53 (15) 4211-4223.

16.   Liu FC, Ma ZY. Achieving exceptionally high superplasticity at high strain rates in a micrograined Al-Mg-Sc alloy produced by friction stir processing. Scripta Materialia 2008; 59 (15) 882-885.

17.   Liu FC, Ma ZY, Chen LQ. Low-temperature superplasticity of Al-Mg-Sc alloy produced by friction stir processing. Scripta Materialia 2009; 60 (5) 968-971.

18.   García-Bernal MA, Mishra RS, Verma R, Hernández-Silva D, High strain rate superplasticity in continuous cast Al-Mg alloys prepared via friction stir processing. Acta Materialia 2009; 60 (10) 850-853.

19. Jana S, Mishra RS, Baumann JB, Grant G. Effect of stress ratio on the fatigue behavior of a friction stir processed cast Al-Si-Mg alloy. Scripta Materialia 2009; 61 (10) 992-995.

20. Gandra J, Miranda RM, Vilaça P. Performance Analysis of Friction Surfacing. Journal of Materials Processing Technology 2012; 212 (8) 1676-1686.

21. Arora HS, Singh H, Dhindaw BK. Composite fabrication using friction stir processing - a review. International Journal of Advanced Manufacturing Technologies 2012; 61 (9-12) 1043–1055.

22. Morisada Y, Fujii H, Nagaoka T, Fukusumim M, MWCNTs/AZ31 surface composites fabricated by friction stir processing. Materials Science and Engineering A 2006; 419 (1-2) 344–348.

23. Dixit M, Newkirk JW, Mishra RS. Properties of friction stir-processed Al1100-NiTi composite. Scripta Materialia 2007; 56 (6) 541-544.

24. Mendes L. Production of aluminium based metal matrix composites reinforced with embedded NiTi by friction stir welding. MSc thesis, Universidade Nova de Lisboa, 2012

25. Shafei-Zarghani A, Kashani-Bozorg SF, Zarei-Hanzaki. Microstructures and mechanical properties of Al/Al2O3 surface nano-composite layer by friction stir processing. Materials Science and Engineering A 2009; 500 (1-2) 84-91.

26. Vitanov VI, Voutchkov II, Bedford GM. Decision support system to optimize the Frictec (friction surfacing) process. Journal of Materials Processing Technologies 2000; 107 (1-3) 236-242.

27. Vitanov VI, Javaid N, Stephenson DJ. Application of response surface methodology for the optimisation of micro friction surfacing process. Surfaces and Coatings Technology 2010; 204 (21-22) 3501-3508.

28. Khalid Rafi H, Janaki Ram GD, Phanikumar G, Prasad Rao K. Microstructural evolution during friction surfacing of tool steel H13. Materials and Design 2011; 32 (1) 82–87.

Chapter 3

A Novel Dynamic Model for Predicting Pressure Wave Velocity in Four-Phase Fluid Flowing along the Drilling Annulus

Xiangwei Kong[1], Yuanhua Lin[2], Yijie Qiu[2], and Xing Qi[2]

[1]School of Chemistry and Chemical Engineering, Daqing Normal University, Daqing, Heilongjiang 163712, China

[2]State Key Laboratory of Oil and Gas Reservoir Geology and Exploitation, Southwest Petroleum University, Chengdu, Sichuan 610500, China

ABSTRACT

A dynamic pressure wave velocity model is presented based on momentum equation, mass-balance equation, equation of state, and small perturbation theory. Simultaneously, the drift model was used to analyze the flow characteristics of oil, gas, water, and drilling fluid multiphase flow. In addition, the dynamic model considers the gas dissolution, virtual mass force, drag force, and relative motion of the interphase as well. Finite difference and Newton-Raphson iterative are introduced to the numerical simulation of the dynamic model. The calculation results indicate that the wave velocity is more sensitive to the increase of gas influx rate than the increase of oil/water influx rate. Wave velocity decreases significantly with the increase of gas influx. Influenced by the pressure drop of four-phase fluid flowing along the annulus, wave velocity tends to increase with respect to well depth, contrary to the gradual reduction of gas void fraction at different depths with the increase of backpressure (BP). Analysis also found that the growth of angular frequency will lead to an increase of wave velocity at low range. Comparison with the calculation results without considering virtual mass force demonstrates that the calculated wave velocity is relatively bigger by using the presented model.

INTRODUCTION

In petroleum industry, managed pressure drilling (MPD) is considered to be one of the most important techniques, which allows accurate control of bottom hole pressure (BHP) by controlling the flow rate, drilling mud density, and back pressure (BP) at the wellhead [1]. As extensively used in the drilling of huge risk, uneconomical, or even abnormal formation, MPD has been of interest in the literature for at least the past decade, especially the related issues about pressure control inducing the dynamic well and kick detection [2]. Due to those efforts, research of the dynamic pressure wave velocity is of great significance to the detection of gas influx and effective control of the pressure at the bottom of well [3]. During the drilling operation of the so-called "microflux control," an MPD technique developed by Santos et al. [4], the return flow is monitored and adjusted to control fluid loss or gain. In the light of control principal, simulation studies were performed to determine the most appropriate initial response to kicks arising due to MPD specific complications caused by BHP fluctuations [5]. During the MPD operations, all unsteady operating, such as changing of pumping rate, adjustment of choke, and controls of BP at wellhead, will generate a pressure wave and threaten the drilling equipment [6].

For the same reason, while tripping out of a drill string in the wellbore, bottom hole is submitted to a suddenly decrease in pressure, leading to fluid expansion and movement out of the annulus. The rapidly expanding fluids and dynamic pressure fluctuations can also lead to rock instability in a reservoir [7]. Particularly, the effects are more important in systems in which multiphase flows occur. New kick-detection tools are now available that are based on acoustic principles, which are of great benefit to potentially earlier and more sensitive detection of a gas influx than pit-gain or paddle flow measurements [8]. The study of propagation of pressure wave is also relevant to control of downhole tool, such as intelligent well downhole control valves applied in different field for many purposes. With further development of oilfield, downhole tool technique for special casing wells is receiving much more attention [9]. At the meantime, an important problem in deep drilling is the propagation of measurement-while-drilling (MWD) mud pulse, transmitting real-time various data from sensors located down hole near the drill bit. Hence, the propagation behavior of pressure wave is considered to provide reference for the MPD operations.

However, we also noticed that the conventional theories present are difficult to be employed in a systematic and accurate prediction when influxes generates. As the influxes fluid including gas, oil, and water will lead to variations of physic characteristics parameter of fluid in the annulus, the distribution of pressure wave velocity in gas, oil, water, and drilling mud four-phase flow along the annulus will be dynamic changing with respect to time and well depth. These effects may influence the safe operation of devices.

Pressure waves are disturbances that transmit energy and momentum along the wellbore through drilling fluid without significant displacement of matter. As migration of dispersed gas, water, and oil towards wellhead is quite complicated, the fundamental characteristics of the four-phase flow are still unknown and modeling results of pressure wave velocity are questionable. The proper treatment of propagation behavior of pressure wave in the four-phase flow in the annulus requires knowledge of description of influxes generation, development, and pressure wave propagation model in a two-phase mixture.

In gas/liquid two-phase flow, two-phase medium interaction greatly changes the structure characteristic of flowing fluid, which results in greater compressibility of two-component flow than single-phase gas or liquid and further causes pressure waves propagating speed to be greatly reduced. This can be explained from the viewpoints of mixture density and compressibility of two-phase fluid and the pressure drop along the flow direction in the wellbore. In the low void fraction range, the gas phase is dispersed in the liquid as bubble, so the wave velocity is influenced greatly by the added gas phase. According to the EOS, if gas invades into the wellbore with a small amount in the bottom hole, the density of the drilling mud has little variation while the compressibility increases obviously, which makes the wave velocity be decreased. This phenomenon was first proposed by Mallock [10] and attracts much attention for its important role in the development of science and technology applications. Extensive investigations involved in the issue of pressure wave propagation have been taken and some significant achievements associated with the theories have also been made in the researches. In early stage researching, Wood [11] extended the researches of Mallock and presented a succinct formula by assuming that the compressibility of two-component fluid is a function related to single-phase compressibility and elastic modulus E. Carstensen and Flody [12] measured the velocity of pressure wave under a lake. A dispersion relation for pressure waves propagating through a bubbly fluid was derived by using a linear scattering theory developed by Foldy. Under the hypothesis of homogeneous and adiabatic laminar flow, Thuraisingham [13] took the two-phase media as a homogeneous fluid and derived the solution model concerning the problem of wave velocity model in two-phase flow at low gas void fraction according to the analysis of state equation of mixture. According to the early stage investigations, Hsieh and Plesset [14] and Murray [15] researched the influence of thermal conductivity and viscosity coefficient on pressure wave propagation. Wallis [16] firstly studied the propagation mechanism of pressure wave and derived the propagation velocity in bubbly flow and separated flow using the homogeneous model. The proposed model is based on the respective compressibility of the vapor and the liquid. Also, the two-phase mixture is treated as a compressible fluid with suitably averaged properties. In that expression, mass and heat transfers are neglected, so it can be applied to any gas/liquid mixture, in so far as no major effect due to vaporization or condensation must be taken into account. It may not be valid for such complex mixtures that never reach equilibrium.

Assuming that no evaporation or condensation occurs when pressure wave is transporting, Moody [17] developed a simple acoustic wave model for bubbly flow and annular flow and established a relationship between sonic velocity and two-phase critical flow. Similar model was developed by D'Arcy [18]. In the researches of Mcwilliam and Duggins [19], surface tension and compressibility of liquid phase were also considered. Henry et al. [20] calculated the velocity as a function of void fraction using a correlation to account for the change in bubble shape with void fraction. Martin and Padmanabhan [21] extended the simple model proposed by Henry by considering wave reflection and wave transmission at gas-liquid interfaces. Researches of Mori et al. [22, 23] suggest that impact pipe elasticity on pressure wave propagation velocity is limited to the range of gas void of less than 1%. At high gas void range, the pressure wave velocity is in between the two velocities of single-phase. Nguyen et al. [24] proposed another type of model for bubbly flow (diluted gas phase in the liquid). The simple relations for prediction of the propagation of pressure disturbances in liquid-gaseous two-phase systems are presented. The model makes use of the well-known physical behavior that the wave velocity in a single-phase fluid is influenced by the elasticity of the confining walls. The interface of the one phase is considered to act as the elastic wall of the other phase and vice versa. Mecredy and Hamilton [25] used the two-fluid model to predict the pressure wave propagation in vapor-liquid flow in detail. However, the analysis contained the important assumption that the evaporation or condensation was governed by kinetic theory. Michaelides and Zissis [26] developed a computational method which yields the sound velocity in terms of the thermodynamic coordinates of the substance without the use of diagrams. Corresponding velocities of sound for the four substances considered exhibit a certain similarity which is examined statistically. The relationship between the sound velocity and the critical mass flux is also investigated. Thuraisingham [13] studied the wave velocity in bubbly water at megahertz frequencies (1~10 MHZ). Numerical analytical results indicate that volume concentrations and the radius of the bubble relative to the incident wavelength of sound are the important parameters which determine the deviation of sound speed form that of bubble-free water.

Currently, the two-fluid continuum model is the most common and reliable method to describe the gas/liquid two-phase flow phenomenon [27]. In the model, the governing equation and phase interface relationship is established based on the assumption that each phase satisfies the continuum conditions. To obtain the practical flow equations, reasonable assumptions and constitutive equations should be introduced. In consequence, the predicted wave velocities were found to depend strongly on the introduced assumptions and equations. In recent years, the two-fluid model was applied in determining the pressure wave propagation characteristics [28]. Ruggles et al. [29, 30] firstly performed the experimental investigation on the dispersion of pressure wave propagation in air-water bubbly flow and studied the propagation of pressure disturbance based on the two-fluid model small perturbation analysis method. Through the comparison with experimental data, they found that the virtual mass force coefficient is a function of gas void. Chung et al. [31] calculated the sonic velocity versus angular frequency form the concept of bubble compressibility in a two-component bubbly flow regime. He also extended such a model to predict the sonic velocity of a vapor-liquid system. Lee et al. [32] constructed the two fluids model to determine the pressure wave propagation speed for two-phase bubbly flow, slug flow, and stratified flow by using pressure disturbance instead of virtual mass and other phase interfacial terms. The results fit well with the experimental in steam water and air water of Henry and theoretical analysis of Nguyen. Zhao and Li [33] derived the general formula of sonic velocity in gas-liquid two-phase flow linear analysis using the linear analysis of the closed fundamental equations of compressible gas-liquid two-phase flow. It is proposed that the appropriate formula for calculating sonic velocity in gas-liquid two-phase flows under usual conditions may be Wood adiabatic sonic velocity formula. By linearizing the conservation equations of two-fluid model, Liu [34] derived a wave number equation of pressure wave for adiabatic gas-liquid two-phase flow. The effects of drag force and virtual mass force on propagation and dispersion of pressure wave were investigated. Xu and Chen [35] used the transient two-fluid model to develop a general relation for acoustic waves with steam-water two-phase mixture in one-dimensional flowing system. Both the mechanical and thermal nonequilibrium are considered. Brennen [36] taken mass and heat exchanges into account and proposed more complete expressions of the speed of sound in two-phase mixture. However, calibration of the mass and heat exchanges requires some further experimental investigations. Yeom and Chang [37] numerically investigated the wave propagation in the two-phase flows. An assessment was made on the effect of interfacial friction terms. Zhang et al. [27] investigated the propagation of the pressure wave in the water-gas two-phase bubbly flow with a one-dimensional two-fluid model and employing small perturbation analysis. The governing equations are simplified and closured according to high-speed aerated flow characteristics in hydraulic engineering. The effects of aerated concentration, liquid pressure, perturbation frequency, and interfacial forces on the acoustic wave velocity and its attenuation in the aerated flow are also explored. With the application of thermal phase change model in computational fluid dynamics code CFX, Li et al. [38] proposed a pressure wave propagation model and investigated the pressure wave propagation characteristics in two-phase fuel systems of liquid-propellant rocket. The propagation of pressure wave during the condensation of R404A and R134A refrigerants in pipe minichannels was given by Kuczyński [39]. Heat exchange between the phases in the condensation process was calculated by using the one-dimensional form of Fourier's equation.

In drilling industry, some scholars have been devoted to this aspect. In the late 1970s, the former Soviet All-Union Drilling Technology Research Institute [40] began to study characteristics of pressure wave velocity in gas-liquid two-phase flow to detect early gas influx and achieved some important results. To study relationship between pressure wave velocity and gas void fraction, Li et al. [41] launched a gas-drilling mud two-phase flow simulation experiment in vertical annulus. It is proved that the two phase flow of gas-liquid patterns and the velocity of gas migration can be determined if the well depth, mud properties, and void fraction in bottom are given The method which is faster in detection time than the method of conventional kick detection was proposed. Starting with the analysis of transient flow, combining with the theory of transmission line, Wang [42] obtained the calculating model of frequency domain for the pulse velocity in drilling fluid built and the impendence and transmission operator of drilling fluid. Alternative initial responses to kicks for various well scenarios during MPD operations were also explored by Davoudi et al. [43]. Li et al. [44] proposed a mathematical model for predicting the attenuation and propagation velocity of measurement while drilling (MWD) pressure pulses in aerated drilling using the two-phase flow model and considering the momentum and energy exchange at the phase interface, gravity of each phase, viscous pipe shear, and other closing conditions. According to the theory of unsteady flow, Xiushan [45] developed the formulas of transmission velocity for mud pulse signal. The formulas which cover all kinds of boundary conditions, including thin wall pipes and thick ones and interaction influence of gas content and solid content on transmission velocity are suitable for positive and negative mud pulse and accord with drilling practice. In a previous work, we proposed a united wave velocity model to predict the pressure wave velocity in gas-drilling mud two-phase steady flow. The effect of well depth, back pressure, gas influx rate, virtual mass force, and angular frequency are all considered. However, under the effects of buoyancy and complicated turbulence interaction, the existing theoretical solutions are not involved in the dynamic model for predicting pressure wave velocity in four-phase fluid flowing along the drilling annulus when influx fluid migrate towards the top of wellbore.

In this paper, the drift model was used to analyze the flow characteristics of oil, gas, water, and drilling fluid multiphase flow. As the important characteristics of influx development, the relative motion of the interphase, such as slippage of gas phase and oil phase, is considered. Moreover, to predict pressure wave velocity in gas-oil-water-drilling mud four-phase flow in the annulus during MPD operations, a dynamic mathematical model is presented. By computing, the influence factors of pressure wave velocity, such as back pressure, gas void fraction, oil void fraction, influx time, influx rate, disturbance angular frequency, and virtual mass force, are analyzed.

MATHEMATICAL MODEL

Before introducing the new dynamic model and to make this point clear, this paper reviews the hydraulic system in MPD operations. The drilling system is a closed circulation with BP at wellhead. The key equipments include the rotating control device, dynamic well control system, conventional pressure control system, industrial personal computer, Coriolis meter, choke, and pressure sensor. First the drilling mud begins to circulate from mud tank, down the drill pipe, and the drill string and returns from the annulus travel back through mud pit where drilling solids are taken away and then to surface mud tank. An important function of the drilling fluid is to provide pressure support to the wellbore wall. The rock formation drilled through has some form of porosity filled with formation fluids. These fluids can be water, or in the case of a reservoir, hydrocarbons. The pressure in these fluids is referred to as the pore pressure. If the pore spaces are connected, these formations will also have permeability. Fluids can flow through them in response to a pressure gradient. The pressure in the annulus is controlled by varying BP to operate the fluid pressure in the wellbore. The aim in MPD is thus to maintain the pressure in the annulus between the two limits of pore pressure and fracture pressure [1].

The United Dynamic Model

When the bottom whole pressure is below the formation pressure, formation fluid will invade into the wellbore and the four-phase flow emerges in the annulus constituted by the drill string and wellbore. As seen in Figure1, take any cross section of the wellbore as an infinitesimal control volume. In the infinitesimal control volume, the four-phase drilling fluid is consisted by drilling mud (considered as a pseudohomogeneous liquid), influx oil (considered as oil phase), influx natural gas (considered as gas phase), and influx water (considered as water phase). Appropriate assumptions and governing equations are critical to simulate realistic four-phase well-control operations. The four-phase model was established based on the following assumptions:

- it is unsteady-state four-phase flow;
- the flow along the flow path is one-dimensional;
- the drilling mud is water-based;
- Drilling mud is incompressible.

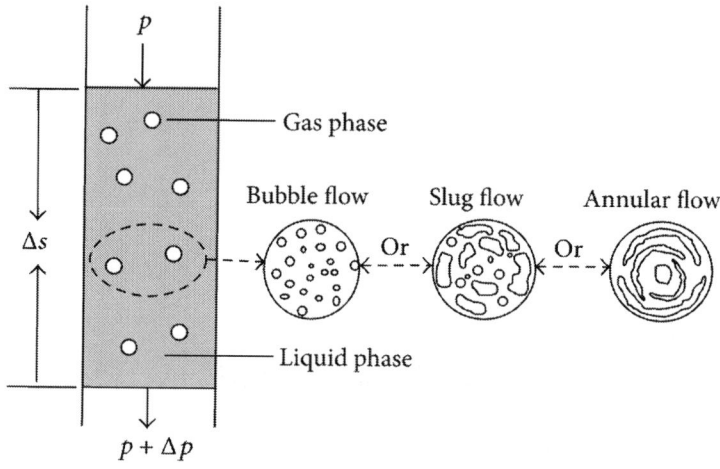

Figure 1: Infinitesimal control volume in effective wellbore.

In the analysis of the multiphase flow characteristic, oil, water, and drilling mud which are all considered as liquid phase have great differences in physical and chemical properties. As water-based medium, water and drilling mud have no substantive difference. In the flow processes, they blend quickly and have no clear phase boundary. Thus, the water-based fluid phase is discussed as water phase, and the oil is considered as another phase.

As the presence of oil and gas interphase mass transfer in the multiphase flow system within the wellbore, the appropriate mass conservation equation can be listed according to oil, gas, and water three components. Given that x_{ik} is the mass fraction of k components in i phase, the mass conservation equations for four-phase mixture are

$$A\frac{\partial}{\partial t}\left(\chi_{ok}\rho_o\phi_o + \chi_{gk}\rho_g\phi_g + \chi_{wi}\rho_w\phi_w + \chi_{mi}\rho_m\phi_m\right)$$

$$+\frac{\partial}{\partial s}\left(\chi_{ok}A\rho_o\phi_o v_o + \chi_{gk}A\rho_g\phi_g v_g\right.$$

$$\left.+\chi_{wi}A\rho_w\phi_w v_w + \chi_{mi}A\rho_m\phi_m v_m\right) = 0. \tag{1}$$

The momentum balance equation for the four-phase mixture is

$$\frac{\partial}{\partial t}\left(A\sum_k \rho_k \phi_k v_k\right) + \frac{\partial}{\partial s}\left(A\sum_k \rho_k \phi_k v_k^2\right)$$

$$+ Ag\sum_k \rho_k \phi_k + \frac{\partial}{\partial s}(Ap) + A\left(\frac{\partial p}{\partial s}\right)_{fr} = 0.$$

(2)

Equation (2) is a general momentum balance equation including hydrostatic pressure gradient, frictional pressure loss gradient, and acceleration loss gradient.

Hence

$$\sum_i \chi_{oi} = 1, \qquad \sum_i \chi_{gi} = 1,$$

$$\sum_i \chi_{wi} = 1, \qquad \sum_i \chi_{mi} = 1,$$

(3)

Where

$$\chi_{wo} = 0, \quad \chi_{wg} = 0, \quad \chi_{ww} = 1, \quad \chi_{wm} = 0,$$

$$\chi_{mo} = 0, \quad \chi_{mg} = 0, \quad \chi_{mw} = 0, \quad \chi_{mm} = 1,$$

$$\chi_{go} = 0, \quad \chi_{gg} = 1, \quad \chi_{gw} = 0, \quad \chi_{gm} = 0,$$

$$\chi_{oo} = \frac{m_o}{m_o + m_g}, \quad \chi_{og} = \frac{m_g}{m_o + m_g}, \quad \chi_{ow} = 0, \quad \chi_{om} = 0.$$

(4)

As propagation velocity is greatly affected by the gas void fraction and angular frequency of the pressure disturbance, the superficial velocity of flowing medium has almost no effect on the propagation velocity [46]; oil, water, and drilling can be considered as liquid phase for their similarity in mechanics. According to the two-fluid model, the flow can be supposed to be gas-liquid two-phase flow from a macroscopic view.

To establish the wave velocity dispersion equation, the mass conservation equations for liquid and gas two phases can be written individually as follows:

$$\frac{\partial}{\partial t}\left(\phi_g \rho_g\right) + \frac{\partial}{\partial s}\left(\phi_g \rho_g v_g\right) = 0,$$

$$\frac{\partial}{\partial t}\left(\phi_L \rho_L\right) + \frac{\partial}{\partial s}\left(\phi_L \rho_L v_L\right) = 0. \tag{5}$$

Hence

$$\rho_L = \phi_o \rho_o + \phi_w \rho_w + \phi_m \rho_m.$$

The gas momentum conservation equation is

$$\frac{\partial}{\partial s}\left(\phi_g \rho_g v_g\right) + \frac{\partial}{\partial s}\left(\phi_g \rho_g v_g^2\right)$$

$$= -\frac{\partial}{\partial s}\left(\phi_g \rho_g\right) + \frac{\partial}{\partial s}\left[\phi_g\left(\tau_g^{\text{fr}} + \tau_g^{\text{Re}}\right)\right] + M_{gi} - 4\frac{\tau_g}{D}. \tag{6}$$

The liquid momentum conservation equation is

$$\frac{\partial}{\partial t}\left(\phi_L \rho_L v_L\right) + \frac{\partial}{\partial s}\left(\phi_L \rho_L v_L^2\right)$$

$$= -\frac{\partial}{\partial s}\left(\phi_L \rho_L\right) + \frac{\partial}{\partial s}\left[\phi_L\left(\tau_L^{\text{fr}} + \tau_L^{\text{Re}}\right)\right] + M_{Li} - 4\frac{\tau_L}{D}. \tag{7}$$

The transfer of momentum M_{gi} and M_{Li} can be written by the following equations:

$$M_{gi} = -M_{Li}^{nd} - M_{Li}^{d} + \left(\tau_{Li}^{fr} + \tau_{Li}^{Re}\right)\frac{\partial\phi_L}{\partial s}$$

$$+ \frac{\partial\left(\phi_g\sigma_s\right)}{\partial s} + \frac{\partial\left(\phi_g P_{gi}\right)}{\partial s} - \phi_g\frac{\partial\left(p_{Li}\right)}{\partial s},$$

$$M_{Li} = M_{Li}^{nd} + M_{Li}^{d} + p_{Li}\frac{\partial\left(\phi_L\right)}{\partial s} - \left(\tau_{Li}^{fr} + \tau_{Li}^{Re}\right)\frac{\partial\phi_L}{\partial s}. \tag{8}$$

Virtual mass force is obtained by the equation in the following form:

$$M_{Li}^{nd} = c_{VM}\phi_g\rho_L\alpha_{VM} - 0.1\phi_g\rho_L v_s\frac{\partial v_s}{\partial s} - c_{m1}\rho_L v_s^2\frac{\partial\phi_g}{\partial s}, \tag{9}$$

where $v_s = v_g$ and $c_{m1} = 0.1$.

The momentum transfer term is described as [47]

$$M_{Li}^{d} = \frac{3}{8}\frac{C_D}{r}\rho_L R_q v_s^2. \tag{10}$$

The pressure difference between the liquid interface and liquid can be obtained by (11)

$$p_{Li} - p_L = -c_p\rho_L v_s^2, \tag{11}$$

Where $c_p = 0.25$

The gas interface pressure P_{gi} is defined as follows:

$$p_{gi} - p_g \approx 0. \tag{12}$$

The pressure of the liquid is

$$p_L = p - 0.25\rho_L\phi_g v_s^2. \tag{13}$$

The shear stress and the interphase shear stress can be described as

$$\tau_g^{fr} \approx \tau_{Li}^{fr} \approx \tau_L^{fr} \approx \tau_g \approx \tau_g^{Re} \approx 0.$$

(14)

The Reynolds stress and interfacial average Reynolds stress are

$$\tau_L^{Re} = -c_r \rho_L v_s^2 \frac{\phi_g}{\phi_L},$$

$$\tau_{Li}^{Re} = -c_r \rho_L v_s^2.$$

(15)

Hence, $c_r = 0.2$.

The wall shear stress of liquid phase is expressed as [48]

$$\tau_L = 0.5 f_L \rho_L v_L^2.$$

(16)

The pressure wave velocity of gas phase c_g and that of liquid phase c_L can be expressed in the following form:

$$\frac{dp_L}{d\rho_L} = c_L^2,$$

$$\frac{dp_g}{d\rho_g} = c_g^2.$$

(17)

Based on (17), the hydrodynamic equations of two-fluid model (5)–(7) can be written in the following matrix form:

$$A\frac{\partial \xi}{\partial t} + B\frac{\partial \xi}{\partial s} = C\xi.$$

(18)

Here, A is the matrix of parameters considered in relation to time; B is the matrix of parameters considered in relation to the spatial coordinate; C is the vector of extractions. By introducing the small disturbance theory, the disturbance of the state variable $\xi(\phi_g, p, V_g, V_L)^T$ can be written as

$$\xi = \xi_0 + \delta\xi \exp\left[i\left(wt - kt\right)\right],\tag{19}$$

where k is the wave number.

According to the solvable condition of the homogenous linear equations that the determinant of the equations is zero, the equation of pressure wave can be expressed in the following form:

$$\begin{vmatrix} \left(\rho_g + c_p\phi_g\rho_1\frac{v_s^2}{c_g^2}\right)w & \frac{\phi_g}{c_g^2}\left[1 - c_p\phi_L\right]\frac{v_s^2}{c_L^2}w & -\left[\phi_g\rho_g k + 2c_p\phi_g\phi_L\rho_L\frac{v_s}{c_L^2}w\right] & 2c_p\phi_g\phi_L\rho_L\frac{v_s}{c_L^2}w \\ -\rho_1 v_1 & \frac{1-\phi_g}{c_L^2}w & 0 & -k\left(1-\phi_g\right)\rho_1 \\ \rho_1 v_1^2 k\left(-\phi_g c_p + c_r - c_i + c_{m2}\right) & -\phi_g k\left[1 - \phi_L\frac{c_p v_s^2}{c_L^2} + c_i\frac{v_s^2}{c_L^2}\right] & \frac{\phi_g\left(\rho_g + c_{vm}\rho_1\right)w}{-i\left(\frac{3}{4}\frac{c_D}{r}\rho_L\phi_g v_s + \frac{4}{D}f_{gw}\rho_g v_g\right)} & -c_{vm}\phi_g\rho_L w + i\left(\frac{3}{4}\frac{c_D}{r}\rho_L\phi_g v_s\right) \\ \rho_L v_s^2 k\left(\phi_L c_p - 2c_i - c_{m2}\right) & -k\left(\phi_L + c_i\phi_g\frac{v_s^2}{c_L^2}\right) & -c_{vm}\phi_g\rho_L w + i\left(\frac{3}{4}\frac{c_D}{r}\rho_L\phi_g v_s\right) & \rho_L\left[\phi_L + \phi_g c_{vm}\right]w \\ & & & -i\left(\frac{3}{4}\frac{c_D}{r}\rho_L\phi_g v_s + \frac{4}{D}f_L\rho_L v_L\right) \end{vmatrix} = 0,\tag{20}$$

where $c_i = 0.3$, $c_{m2} = 0.1$, $c_r = 0.2$, and $c_p = 0.25$.

The real value of wave number is determined as the pressure wave velocity c, and pressure wave velocity in the four-phase flow is

$$c = \frac{\left|w/R^+\left(k\right) - w/R^-\left(k\right)\right|}{2}.\tag{21}$$

Physical Models

To define the velocity of pressure wave propagation in the four-phase flow, the related physical models are required, such as equations of state, temperature distribution model, gas dissolution, and oil phase volume factor.

Equations of State for Gas

The equation of state (EOS) for gas can be expressed as

$$c = \frac{\left|w/R^+\left(k\right) - w/R^-\left(k\right)\right|}{2}.\tag{22}$$

For p<35 MPa, the compression factor is obtained as follows:

$$Z_g = 1 + \left(0.3051 - \frac{1.0467}{T_r} - \frac{0.5783}{T_r^3} \right) \rho_r$$

$$+ \left(0.5353 - \frac{02.6123}{T_r} - \frac{0.6816}{T_r^3} \right) \rho_r^2, \tag{23}$$

where $T_r = T/T_c$, $p_r = p/p_c$, $\rho_r = 0.27 p_r / Z_G T_r$.

For p≥35 MPa, the compression factor under the condition of high pressure is [49]

$$Z_g = \frac{0.06125 P_r T_r^{-1} \exp\left(-1.2\left(1 - T_r^{-1}\right)^2\right)}{Y}, \tag{24}$$

where Y is given by the follow equations:

$$- 0.06125 P_r T_r^{-1} \exp\left(-1.2\left(1 - T_r^{-1}\right)^2\right) + \frac{Y + Y^2 + Y^3 + Y^4}{(1 - Y)^3}$$

$$= \left(14.76 T_r^{-1} - 9.76 T_r^{-2} + 4.58 T_r^{-3}\right) Y^2$$

$$- \left(90.7 T_r^{-1} - 242.2 T_r^{-2} + 42.4 T_r^{-3}\right) Y^{(2.18 + 2.82 T_r^{-1})}. \tag{25}$$

Equations of State for Liquid

With T<130°C, the density of drilling mud is expressed as follows:

$$\rho_m = \rho_0 \left(1 + 4 \times 10^{-10} p_L - 4 \times 10^{-5} T - 3 \times 10^{-6} T^2\right) \tag{26}$$

With T≥130°C, the density of drilling mud is expressed as follows:

$$\rho_m = \rho_0 \left(1 + 4 \times 10^{-10} p_L - 4 \times 10^{-5} T\right.$$

$$\left. -3 \times 10^{-6} T^2 + 0.4 \left(\frac{T - 130}{T}\right)^2\right) \tag{27}$$

Temperature Distribution Model

The temperature of the drilling mud at any depth of the wellbore is [50]

$$T = T_{ei} + F\left[1 - e^{(z_{bh}-z)/A}\right]\left(-\frac{g\sin\theta}{g_c J c_{pm}} + N + g_T\sin\theta\right)$$
$$+ e^{(z_{bh}-z)/A}\left(T_{fbh} - T_{ebh}\right).$$

$$(28)$$

Gas Dissolution

Assuming gas goes into and comes out of solution instantaneously, gas solubility can be obtained by (29)

$$0.021\gamma_{gs}\left[(p + 0.1757)\,10^{(1.7688/\gamma_{os}-0.001638T)}\right]$$

$$(29)$$

Oil Phase Volume Factor

The volume factor is calculated as follows:

$$= 0.976 + 0.00012\left[5.612\left(\frac{\gamma_{gs}}{\gamma_{os}}\right)^{0.5}R_s + 2.25T + 40\right]$$

$$(30)$$

Flow Pattern Prediction Models

The flow regime, the flow pattern, and structure of the flow are some of the important parameters to describe two-phase gas-quid flows, identify two phase gas-quid flow regimes, and calculate the dynamic pressure wave propagation velocity. Thus, the transition among the three main flow regimes (bubbly, slug, and annular) is desirable to be known [51].

In the hydraulic calculation of annulus, the effective diameter of annulus [52] should be given as

$$D = \frac{\pi\left(D_o^2 - D_i^2\right)/4}{\pi\left(D_o + D_i\right)/4} = D_o - D_i.$$

$$(31)$$

The effective roughness of annulus can be calculated by

$$k_e = k_0 \frac{D_o}{D_o + D_i} + k_i \frac{D_i}{D_o + D_i}.$$

(32)

At low gas flow rate, the liquid is continuous phase, and the gas bubbles are dispersed in the liquid phase. Studies of Taitel [53] give the minimum diameter necessary to form bubbly flow as

$$D_{min} = 19 \left[\frac{\left(\rho_L - \rho_g \right) \sigma_s}{g \rho_L^2} \right]^{0.5}$$

(33)

The critical condition for forming bubbly flow is

$$v_{Mcr}^{1.12} = 5.88 D^{0.48} \left[\frac{g \left(\rho_L - \rho_g \right)}{\sigma_s} \right]^{0.5} \left(\frac{\sigma_s}{\rho_L} \right) \left(\frac{\rho_M}{\mu_L} \right)^{0.08},$$

(34)

$$D > D_{min},$$

$$\phi_g \leq 0.25, \qquad v_M \leq v_{Mcr},$$

$$\phi_g \leq 0.52, \qquad v_M > v_{Mcr}.$$

(35)

For slug flow, the critical balance superficial flow rate of gas carrying droplets needs to meet the condition [54] that

$$\left[v_{sg} \right]_{cr} = 3.1 \left[\frac{g \sigma \left(\rho_L - \rho_g \right)}{\rho_g^2} \right]^{0.25},$$

(36)

$$\phi_g > 0.25, \quad v_M \leq v_{Mcr},$$

$$\phi_g > 0.52, \quad v_M > v_{Mcr},$$

$$v_{sg} \leq [v_{sg}]_{cr}.$$

$$(37)$$

For annular flow, the pattern transition criterions [55] is

$$v_{sg} > [v_{sg}]_{cr}.$$

$$(38)$$

Bubbly Flow

Gas void fraction of four-phase flow is described as

$$\phi_g = \frac{v_{sg}}{S_g \left(v_{so} + v_{sg} + v_{sw} + v_{sm} \right) + v_{gr}}.$$

$$(39)$$

The value of the distribution factor S_g can be determined by

$$S_g = 1.20 + 0.371 \left(\frac{D_i}{D_o} \right).$$

$$(40)$$

Harmathy [56] established the calculation formula of gas slip velocity in bubbly flow based on the study of the migration velocity of the bubble in a stationary liquid as

$$v_{gr} = 1.53 \left[\frac{g\sigma_s \left(\rho_L - \rho_g \right)}{\rho_L^2} \right]^{0.25}.$$

$$(41)$$

The average density of four-phase mixture flow is

$$\rho_M = \phi_L \rho_L + \phi_g \rho_g.$$

$$(42)$$

The oil void fraction for four-phase flow is

$$\phi_o = \frac{\left(1 - \phi_g\right) v_{so}}{S_o \left(v_{so} + v_{sw} + v_{sm}\right) + \left(1 - \phi_g\right) v_{or}}.$$

(43)

The value of the distribution factor is $S_o = 1.05 + 0.371(D_i/D_o)$

On the basis of the total liquid fluid, establish the oil phase velocity relationship as

$$v_o = S_o v_L + v_{or}.$$

(44)

According to cross-section flow rate, phase distribution, and slip mechanism of liquid phase, we can the draw the following relationship:

$$v_{or} = 1.53 \left[\frac{g\sigma_{wo} - \rho_o}{\rho_{wb}^2} \right]^2,$$

(45)

Where

$$\rho_{wb} = \phi_w \rho_w + \phi_m \rho_m.$$

(46)

Water void fraction is

$$\phi_w = \frac{\left(1 - \phi_g - \phi_o\right) v_{sw}}{v_{sw} + v_{sm}}.$$

(47)

Drilling mud void fraction is

$$\phi_m = 1 - \phi_g - \phi_o - \phi_w.$$

(48)

Due to the similar physical properties of water and drilling fluids, v_w, v_m, and v_{wb} can be expressed by

$$v_w = v_m = v_{wb}.$$

(49)

The coefficient of virtual mass force C_{vm} for bubbly flow can be expressed as follows:

$$C_{vm} = 0.5 \frac{1 + 2\phi_g}{1 - \phi_g}.$$

(50)

The coefficient of resistance coefficient C_D for bubbly flow can be expressed by

$$C_D = \frac{4R_b}{3} \sqrt{\frac{g(\rho_L - \rho_g)}{\sigma_s}} \left[\frac{1 + 17.67\phi_L^{9/7}}{18.67\phi_L^{1.5}} \right]^2.$$

(51)

The friction pressure gradient for bubbly flow can be obtained from the following equation:

$$\tau_f = f \frac{\rho_L v_L^2}{2D}.$$

(52)

Slug Flow

The void fraction of the four phases Φ_g, Φ_o, Φ_w and Φ_m can be determined by (39), (43), (47), and (48) the same as bubbly flow.

The value of the distribution factor S_g for slug flow can be described as

$$S_g = 1.182 + 0.9 \left(\frac{D_i}{D_o} \right).$$

(53)

For slug flow, the slip velocity can be calculated as follows.

From experimental studies, Hasan and Kabir [57] established the calculation formula of drift velocity for slug flow on the basis of research on Taylor bubble migration rule of Davies and Taylor as

$$v_{gr} = \left(0.35 + 0.1\frac{D_i}{D_o} \right) \left[\frac{gD_o(\rho_L - \rho_g)}{\rho_L} \right]^{0.5}.$$

(54)

The coefficient of virtual mass force C_{vm} for slug flow can be expressed as follows:

$$C_{vm} = 3.3 + 1.7 \frac{3I_q - 3R_q}{3L_q - R_q}.$$

(55)

The coefficient of resistance coefficient C_D for slug flow can be expressed as

$$C_D = 110 \phi_L^3 R_b.$$

(56)

Annular Flow

As for annular flow, due to the miscible flow state of gas at center, the simplification can be .$V_{gr} = 0$

The void fraction of gas can be determined by

$$\phi_g = \left(1 + X^{0.8}\right)^{-0.378},$$

(57)

where X is defined as

$$X = \sqrt{\frac{(dp/ds_L)_{fr}}{(dp/ds_g)_{fr}}}.$$

(58)

Oil void fraction is

$$\phi_o = \frac{\left(1 - \phi_g\right) v_{so}}{v_{so} + v_{sw} + v_{sm}}.$$

(59)

Water void fraction is

$$\phi_w = \frac{\left(1 - \phi_g\right) v_{sw}}{v_{so} + v_{sw} + v_{sm}},$$

$$\phi_m = 1 - \phi_g - \phi_o - \phi_w.$$

(60)

The same as slug flow, C_{vm} and C_D can be determined by (55) and (56).

SOLUTION OF THE DYNAMIC MODEL

Now since obtaining the analytic solution of the aforementioned theoretical model directly is impossible, discretization of the model to a numerical model is required [58]. In this paper, the mathematical methods based on finite difference method provide a numerical solution approach for the dynamic model. As for the solution of the pressure wave velocity model, spatial domain includes the entire wellbore and the formation node; time domain is the time period, influx fluid flowing from the bottom hole to the wellhead along the wellbore. Discretizing the domain of determinacy, the entire spatial and time domain can be divided into discrete networked systems.

According to finite difference scheme, the four equations (5)–(7) are solved by using the finite difference method with computational cells shown in Figure 2; the difference equation systems of which described the basic principles of four-phase fluid motion in wellbore is presented as follows.

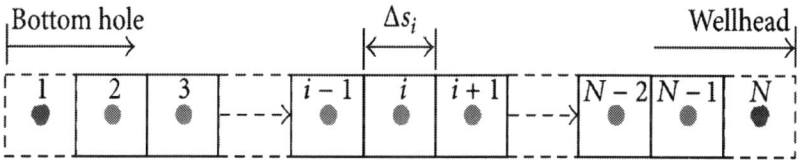

Figure 2: Computational cells for semi-implicit difference solution.

For the drilling mud phase,

$$\frac{\left(Av_{sm}\right)_{i+1}^{n+1} - \left(Av_{sm}\right)_{i}^{n+1}}{\Delta s}$$

$$= \frac{\left(A\phi_m\right)_i^n + \left(A\phi_m\right)_{i+1}^n - \left(A\phi_m\right)_i^{n+1} - \left(A\phi_m\right)_{i+1}^{n+1}}{2\Delta t}. \tag{61}$$

For the water phase,

$$\frac{\left(Av_{sw}\right)_{i+1}^{n+1} - \left(Av_{sw}\right)_i^{n+1}}{\Delta s}$$

$$= \frac{\left(A\phi_w\right)_i^n + \left(A\phi_w\right)_{i+1}^n - \left(A\phi_w\right)_i^{n+1} - \left(A\phi_w\right)_{i+1}^{n+1}}{2\Delta t}. \tag{62}$$

For the oil phase,

$$\frac{\left(A\left(v_{so}/B_o\right)\right)_{i+1}^{n+1} - \left(A\left(v_{so}/B_o\right)\right)_i^{n+1}}{\Delta s}$$

$$= \left(\left(A\frac{\phi_{so}}{B_o}\right)_i^n + \left(A\frac{\phi_{so}}{B_o}\right)_{i+1}^n - \left(A\frac{\phi_{so}}{B_o}\right)_i^{n+1} \right.$$

$$\left. - \left(A\frac{\phi_{so}}{B_o}\right)_{i+1}^{n+1} \right) \times (2\Delta t)^{-1}. \tag{63}$$

For the gas phase,

$$\frac{\left[A\left(\rho_g v_{sg} + \rho_{gs}R_s v_{so}/B_o\right)\right]_{i+1}^{n+1} - \left[A\left(\rho_g v_{sg} + \rho_{gs}R_s v_{so}/B_o\right)\right]_i^{n+1}}{\Delta s}$$

$$= \frac{\left[A\left(\rho_g\phi_g + \rho_{gs}R_s\phi_o/B_o\right)\right]_i^n}{2\Delta t}$$

$$+ \frac{\left[A\left(\rho_g\phi_g + \rho_{gs}R_s\phi_o/B_o\right)\right]_{i+1}^n}{2\Delta t}$$

$$- \frac{\left[A\left(\rho_g\phi_g + \rho_{gs}R_s\phi_o/B_o\right)\right]_i^{n+1}}{2\Delta t}$$

$$- \frac{\left[A\left(\rho_g\phi_g + \left(\rho_{gs}R_s\phi_o/B_o\right)\right)\right]_{i+1}^{n+1}}{2\Delta t}. \tag{64}$$

For momentum balance equation

$$\left(Ap\right)_{i+1}^{n+1} - \left(Ap\right)_i^{n+1} = \xi_1 + \xi_2 + \xi_3 + \xi_4, \tag{65}$$

where

$$\xi_1 = \frac{\Delta s}{2\Delta t}\left[\begin{array}{c} \left(A\left(\rho_m v_{sm} + \rho_w v_{sw} + \rho_o v_{so} + \rho_g v_{sg}\right)\right)_i^n + \left(A\left(\rho_m v_{sm} + \rho_w v_{sw} + \rho_o v_{so} + \rho_g v_{sg}\right)\right)_{i+1}^n \\ -\left(A\left(\rho_m v_{sm} + \rho_w v_{sw} + \rho_o v_{so} + \rho_g v_{sg}\right)\right)_i^{n+1} - \left(A\left(\rho_m v_{sm} + \rho_w v_{sw} + \rho_o v_{so} + \rho_g v_{sg}\right)\right)_{i+1}^{n+1} \end{array}\right], \tag{66}$$

and

$$\xi_2 = \left[A\left(\frac{\rho_m v_{sm}^2}{\phi_m} + \frac{\rho_w v_{sw}^2}{\phi_w} + \frac{\rho_g v_{sg}^2}{\phi_g} + \frac{\rho_o v_{so}^2}{\phi_o}\right)\right]_i^{n+1}$$

$$- \left[A\left(\frac{\rho_m v_{sm}^2}{\phi_m} + \frac{\rho_w v_{sw}^2}{\phi_w} + \frac{\rho_g v_{sg}^2}{\phi_g} + \frac{\rho_o v_{so}^2}{\phi_o}\right)\right]_{i+1}^{n+1},$$

$$\xi_3 = -\frac{g\Delta s}{2}\left[(A\rho_M)_i^{n+1} + (A\rho_M)_{i+1}^{n+1}\right],$$

$$\xi_4 = -\frac{\Delta s}{2}\left[\left(A\left(\frac{\partial p}{\partial s}\right)\right)_{fri}^{n+1} + \left(A\left(\frac{\partial p}{\partial s}\right)\right)_{fri+1}^{n+1}\right]. \tag{67}$$

Besides finite-difference method, the characterized method and the Newton-Raphson iterative method are adopted to solve the united dynamic model in four phases along annulus. On the basis of discretization, the solution of models was realized by applying personally complied code on VB. NET and the solution procedure of that are shown in Figure 3.

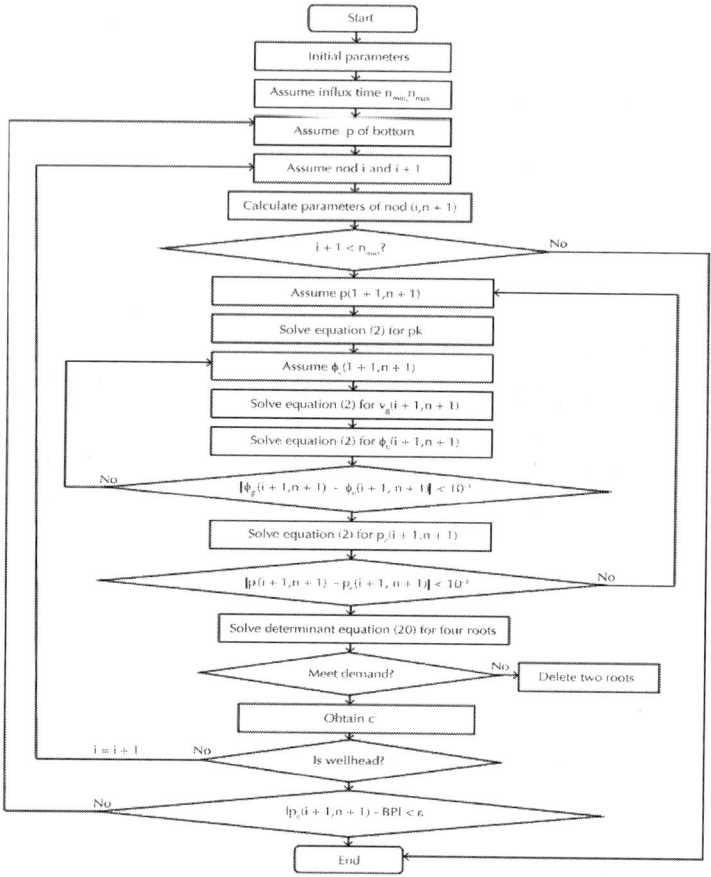

Figure 3: Solution procedures for pressure wave velocity in MPD operations.

The four-phase flow system is described completely by eight variables including pressure, temperature, gas void fraction and liquid void fraction, gas and liquid densities, and gas and liquid velocities. There are still five unknowns, such as gas and liquid velocities, gas fraction, pressure, and gas density based on the above assumptions. Therefore, five equations are required to compute the unknown variables with boundary conditions. At different annulus depth, we can obtain pressure, temperature, gas velocity, oil velocity, water velocity, drilling mud velocity and gas void fraction, oil void fraction, and water void fraction during MPD operations by application of the finite-difference. At initial time, the wellhead BP, wellbore structure, well depths, wellhead temperature, gas-oil-water-drilling mud properties, and so forth are known. On node i, flow parameters, such as pressure, temperature, and void fraction of gas-oil-water-drilling mud multiphase, can be obtained by adopting finite difference. Then, the determinant equation (20) is calculated based on the calculated parameters. Omitting the two unreasonable roots, the pressure wave at different depth of annulus in MPD operations can be solved by (21). Upon substitution of the actual magnitudes, v turned out to be the velocity of light. The process is repeated until the pressure wave velocity in the whole annulus has been obtained.

The published experimental data presented by Li et al. [41] is used to verify the new developed united dynamic model. In Figure 4, the points represent Li's experimental data and the lines represent the calculation results. From the comparisons, it can be concluded that the developed united dynamic model agrees well with the experimental data. The averaged error between model and data in Figure 4 is 2.853%.

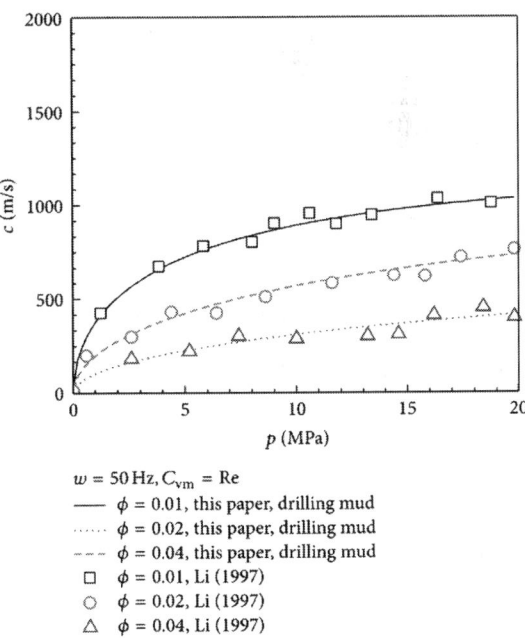

$w = 50\,\text{Hz}, C_{vm} = \text{Re}$
—— $\phi = 0.01$, this paper, drilling mud
······ $\phi = 0.02$, this paper, drilling mud
--- $\phi = 0.04$, this paper, drilling mud
□ $\phi = 0.01$, Li (1997)
○ $\phi = 0.02$, Li (1997)
△ $\phi = 0.04$, Li (1997)

Figure 4: Comparison between calculation results of new dynamic model and the experimental data of Li.

ANALYSIS AND DISCUSSION

One of the potentially most useful capabilities of transient well control model is the ability to simulate various choke control procedures. Typically, the control of choke to maintain constant BHP depends on the experience and training of the hand on the choke. The gas and liquid flow rate measured by Coriolis meter and the BP measured by pressure senor is taken as the initial data for annulus pressure calculation. The well used for calculation is a gas well in Sichuan Region, China. The wellbore structure of this well is shown in Figure 5, and information about gas-oil-water-drilling mud properties, well design parameters, and operational conditions of calculation well are listed in Table 1.

Table 1: Parameters of the calculation well

Type	Property	Value
Mud	Density (kg/m³)	1460
	Viscosity (Pa·s)	20×10^{-3}
Gas	Relative density	0.65
	Viscosity (Pa·s)	1.14×10^{-5}
Oil	Density (kg/m³)	800
	Viscosity (Pa·s)	40×10^{-3}
Water	Density (kg/m³)	1000
	Viscosity (Pa·s)	1.0×10^{-3}
String	Elastic modulus of string (Pa)	2.07×10^{11}
	Poisson ratio of string	0.3
	Roughness (m)	1.54×10^{-7}
Surface condition	Surface temperature (K)	298
	Atmosphere pressure (MPa)	0.101

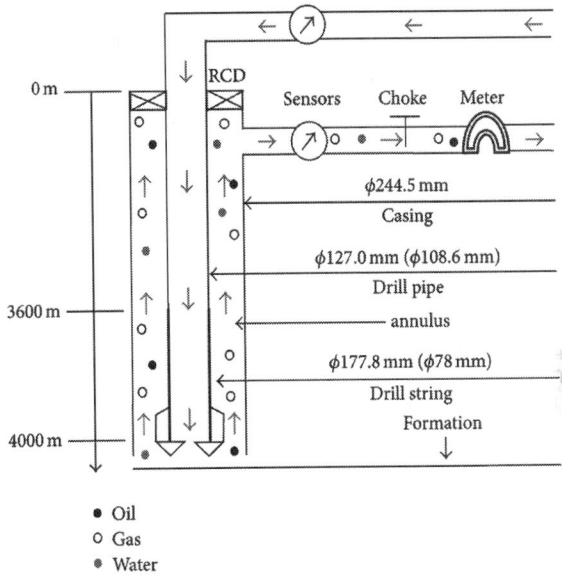

Figure 5: Structural parameters of the calculation well.

In the following example, the drilling mud mixed with gas, oil, and water is taken as a four-phase flow medium when the influx of gas, oil, and water occurs at the bottom of the well. In the calculation, the BP is 0.2MPa, and flow rate is Q_g= 3.6m³/h, Q_w= 3.6m³/h, Q_o= 3.6m³/h, and Q_m=72.0m³/h, respectively. The propagation velocity of pressure wave in the four-phase flow is calculated and discussed by using the established model and well parameters.

Effect of Well Depth on the Pressure Wave Velocity

With bottom hole pressure being constant, the effect of well depth on pressure wave velocity is analyzed. When gas-oil-water influx fluid invades into the wellbore at the bottom hole, gas, oil, water, and drilling mud four-phase flow appears in the annulus and migrates from the bottom of the well (H=4000 m) to the wellhead along the flow direction. It is assumed that the initial length of the gas pillar is 500 m under the bottom hole pressure and temperature condition in the analysis of the migration of influx gas. The migration and change of gas column including the solubility and length change lead to the variation of pressure wave velocity in the annulus. The formation and development of influx are a dynamic process, and the flow of fluid in the annulus is very complicated. The flow condition and flow parameter at any time are various.

Figure 6 illustrates the distribution of pressure wave velocity at different well depth after 5.16 minutes and 18.71 minutes' development of influx. It is shown that the top of aerated mud column arrives at the position of H=3000 m and H=2000 m, respectively. Due to the circulation of drilling mud and slip of gas phase, the length of aerated mud column that represents the rising height of gas will increase with the gradual decrease of pressure along the annulus. Besides, in a segment of aerated mud column, the gas void fraction of the top is higher than that of the bottom for the slop of gas phase. So, in the declining stage, pressure wave velocity in aerated mud column presents an increasing tendency along the flow direction. As the gas migrates upwards, the casing pressure should be changed dynamically to keep the bottom hole pressure constant, which will lead to an increase of annular pressure and back pressure. As the influx develops for 18.71 minutes, the back pressure will be prompted to 2.39 MPa, increased by 0.29 MPa compared with that of 5.16 minutes.

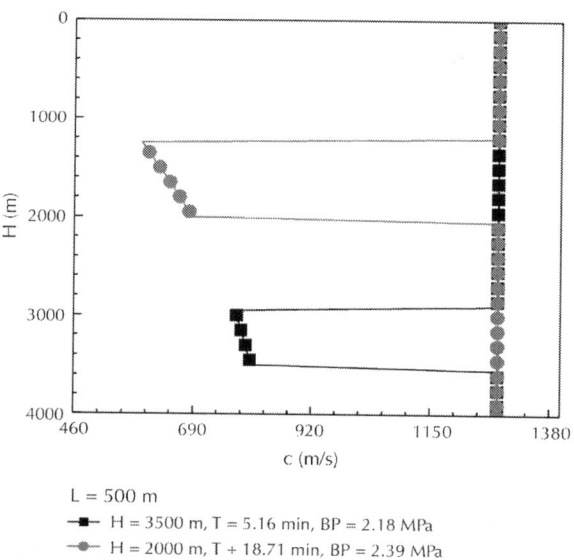

L = 500 m
-■- H = 3500 m, T = 5.16 min, BP = 2.18 MPa
-●- H = 2000 m, T + 18.71 min, BP = 2.39 MPa

Figure 6: Wave velocity variations at different depth.

Effect of BP on Pressure Wave Velocity

In Figures 7 and 8, the three-dimensional figures illustrate the distributions of gas void fraction and variations of pressure wave velocity along the flow direction in the annulus at different time when BP is 0.1 MPa, 3.0 MPa, 6.0 MPa, and 9.0 MPa, respectively. When influx occurs, increase of applied BP is of the most commonly used approach for bottom hole pressure control. Control of back pressure has a great influence on the gas void fraction. In the two graphs, the two full lines give the shape of the void fraction and pressure wave velocity at different back pressure. In each graph, the two clusters of dotted lines at the two-dimensional plane are the projection of the full lines, which, respectively, represent the influence of well depth and time on the distributions of gas void fraction or variations of pressure wave velocity at different back pressure. Increasing the casing pressure will make the compressibility decrease obviously and energy dissipation with the propagation of pressure wave decrease as well. Thus, it can be seen that, with the reduction of back pressure, void fraction shows an increasing trend and that the pressure wave velocity has a decrease at different depth of well. Meanwhile, at the high temperature and high pressure bottom hole, pressure wave velocity change in a comparatively smooth level in a consequence of compressibility of the four-phase fluid changing only in a small degree. As a result, as aerated drilling mud migrating from the bottom hole to the wellhead, gas void fraction significantly increases first slightly and then sharply along the flow direction a drop of annular pressure. Conversely, the pressure wave velocity shows a gradual fist and then remarkable decreasing tendency after the influx gas reaches the position close to the wellhead. Also, with the development of influx, the influence of BP on the gas void fraction and pressure wave velocity is larger over time.

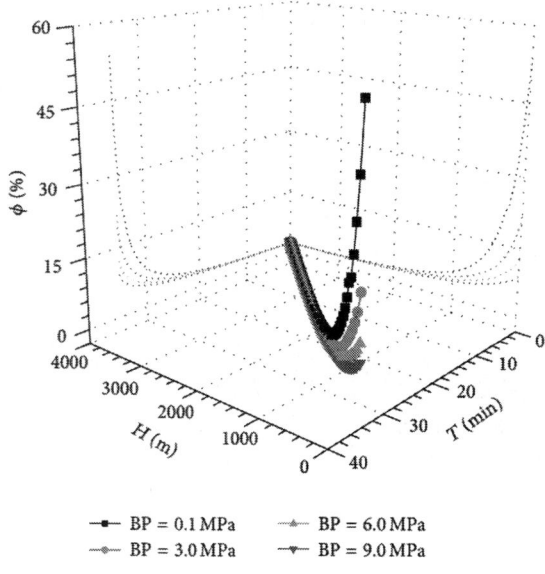

Figure 7: Effect of BP on the gas void fraction.

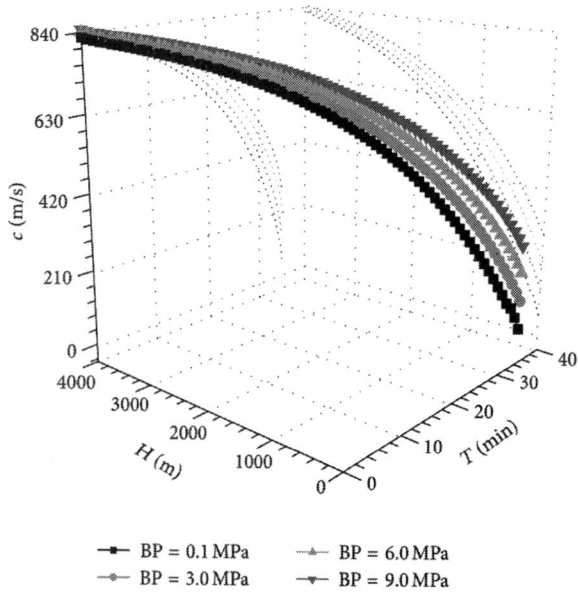

Figure 8: Effect of BP on the pressure wave velocity.

Effect of Gas Influx Rate on Pressure Wave Velocity

Figures 9 and 10 graphically interpret the effect of gas influx rate on the distribution of gas void fraction and variations of pressure wave velocity along the flow direction in the annulus. The same as Figures 7 and 8, the two clusters of dotted lines at the two-dimensional plane are the projection of the full lines. The pressure wave velocity at different well depth is the transient value with respect to time. It is noted that the changes of pressure wave velocity are obvious with the migration of gas, which is extremely significant at wellhead even at low gas influx rate. With the decreasing of pressure at the position near the wellhead, rapid expansion of gas volume appears, and as a result the gas void fraction increases sharply and the pressure wave velocity decrease obviously at the same time. However, the gas void fraction and pressure wave velocity have no obvious change at the bottom of well due to the high pressure at the bottom of well and insignificant compressibility of gas. When the mainly reflect fundamental factors, initial gas influx rate, is increased, pressure wave velocity shows an obvious descending trend with the increase of initial gas influx rate, the value of pressure wave velocity at the position close to the wellhead is much greater than that at bottom hole. At the gas void fraction range of about 5%~80%, as gas migrating along the annulus, the pressure wave velocity is decreased to a minimum at the wellhead with the gas void fraction increasing to a maximum.

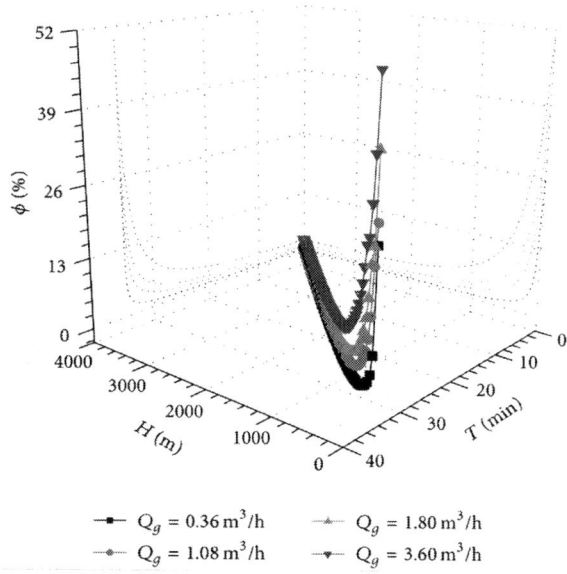

Figure 9: Gas void fraction distribution at different gas influx rate.

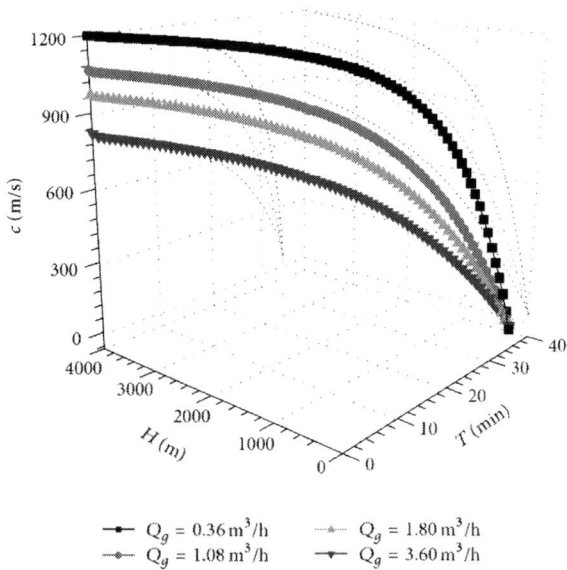

Figure 10: Wave velocity variations at different gas influx rate.

Effect of Oil Influx Rate on Pressure Wave Velocity

Figure 11 shows the effect of oil influx on the gas void fraction in gas-oil-water-drilling mud four-phase flow. The increase of oil influx rate has little influence on the gas void fraction distribution, which is because gas void fraction is mainly impacted by pressure, temperature, and gas influx rate. When the oil influx rate increased from Q_g=0.36m³/h to Q_g=3.60m³/h, the change of mixture density is inconspicuous compared with drilling mud. Therefore, the gas void fraction keeps almost unchanged. Similarly, as shown in Figure 12, with the oil influx rate increase, the change of pressure wave velocity can be neglected due to the little influence of oil influx rate on the gas void fraction. Also, when the water influx invades the wellbore, the influx water will dissolve in drilling mud and reduce the density of drilling mud as the water phase and water base drilling mud have similar physical properties. But the influence of changing water influx rate on pressure wave velocity is too little to be neglected as the oil influx rate does.

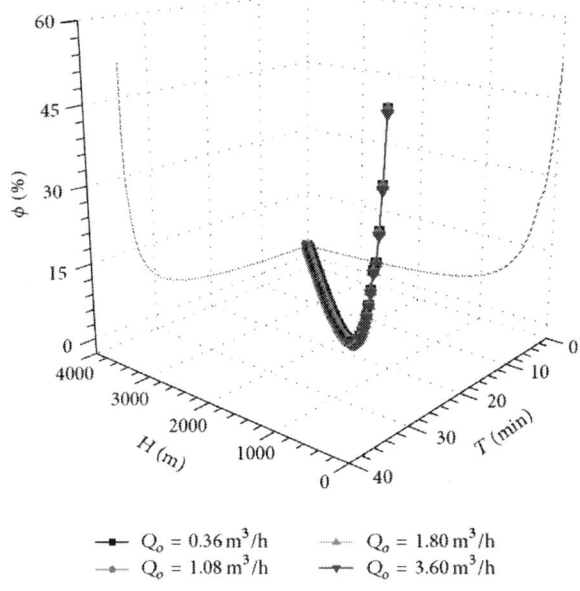

Q_o = 0.36 m³/h Q_o = 1.80 m³/h
Q_o = 1.08 m³/h Q_o = 3.60 m³/h

Figure 11: Gas void fraction distribution at different oil influx rates.

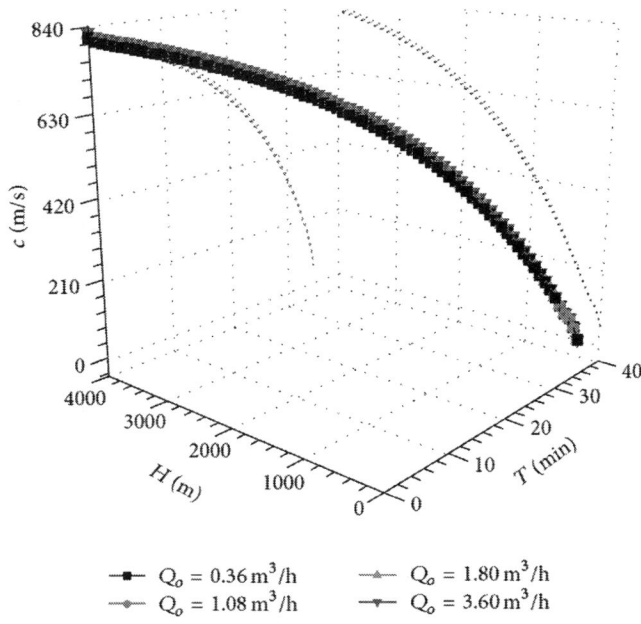

Figure 12: Wave velocity variations at different oil influx rates.

Effect of Disturbance Angular Frequency on Pressure Wave Velocity

Figure 13 provides an illustration of the influence of disturbance angular frequency on the pressure wave velocity when gas, oil, and water influx generates. It is indicated that an increase in angular frequency results in a rise of pressure wave velocity. In addition, the influence on dynamic pressure wave velocity is enhanced with the continuation of time and the decrease of well depth as the gas migrates from the bottom hole to the wellhead. Accordingly, the disturbance angular frequency has a significantly distinct influence on the pressure wave velocity at the position near the wellhead due to the phenomenon that sharply increases of gas void fraction. For an increase in angular frequency from 1500 Hz to 10000 Hz, the influence on wave velocity is not as great as the increase form 50 Hz to 1500 Hz. This is consistent with the fact that the effect of angular frequency on the pressure wave appears at a low angular frequency.

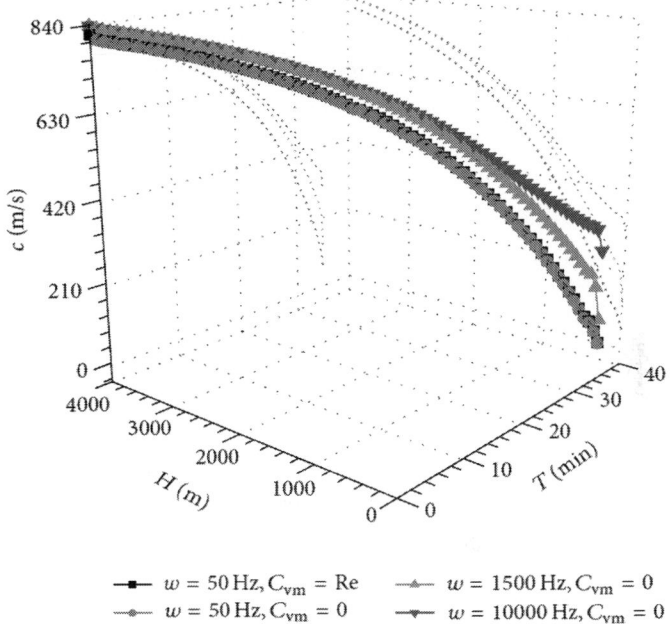

Figure 13: Pressure wave velocity variations at different angular frequency.

CONCLUSIONS

Solution of dynamic pressure wave velocity in gas-oil-water-drilling mud four-phase flow is a lot more complicated than that in two-phase flow. The dynamic model which takes consideration of gas solubility, oil compression coefficients, perturbation frequency, virtual mass force, and drag force is established. Owing to the solution of the migration of drilling mud which contains influx fluid (gas, water, and oil), the dynamic pressure wave varied with time and well depth is solved, and the main conclusions can be summarized as follows.

- Agreeing well with the previous experimental data, the developed untied dynamic model can be used to calculate the wave velocity in the annulus. The model can accurately give the dynamic at different depth and time during the migration of influx, which will be beneficial to provide a reference for well control and MWD.

- In the dynamic migration process, the pressure wave velocities at different well depth are changing constantly due to the circulation of drilling mud and slip of gas phase. The length of aerated mud column which represents the rising height of gas increases with the gradually decreasing pressure along the annulus. In the segment of aerated mud column, pressure wave velocity presents an increasing tendency along the flow direction.

- The flow condition and flow parameter at any time are various. Both the static pressure and void fraction distribution along the annulus changed with the position in the annulus and time and have direct impact on the pressure wave velocity. The pressure wave velocity happens to decrease after an elapsed period of time in which influx fluid migrates along the annulus. The decreasing tendency will last until the aerated mud column pass through that point of annulus completely. Also, the time required is gradually shortened during process of influx migration and growth.

- At the bottom hole, the void of gas, water, and oil is very low relative to drilling mud. With the migration of influx fluid, the variation of gas void fraction is much greater than other phases; in particular, a sharp increase of gas void fraction presents at the position near the wellhead for the rapid expansion of gas volume, while the void fractions of water, oil, and drilling mud are decreased for the nearly negligible variations of physical properties and drastic expansion of gas at the position near the wellhead.

- When the BP at the wellhead increases, gas void fraction at any position of annulus decreases; instead, the pressure wave velocity increases. Moreover, the influence of BP on the gas void fraction is larger at position close to the wellhead than that at the bottom of well. With the development of influx, the influence of BP on the gas void fraction and pressure wave velocity is larger over time.

- The pressure wave velocity is sensitive to the gas influx rate, in particular at the position close to the wellhead. Even at low range, the rise of gas influx rate also will result in a significantly downward trend of pressure wave velocity. On the contrary, the influence of changing water influx rate and oil influx rate on pressure wave velocity is too little to neglect.

- Ignoring virtual mass force, the pressure wave velocity has an obvious dispersion characteristic at different position of well. The effect of angular frequency on the pressure wave appears at a low angular frequency, and the pressure wave velocity increases together with the growth of the angular frequency.

ACKNOWLEDGMENTS

This research work was cofinanced by the National Natural Science Foundation of China (no. 51274170) and Important National Science & Technology Specific Projects (2011ZX05022-005-005HZ). Without their support, this work would not have been possible.

REFERENCES

1.   W. A. Bacon, Consideration of compressibility effects for applied-back- pressure dynamic well control response to a gas kick in managed pressure drilling operations [M.S. thesis], University of Texas, Arlington, Tex, USA, 2011.

2.   S. Saeed and R. Lovorn, "Automated drilling systems for MPD-the reality," in Proceedings of the IADC/SPE Drilling Conference and Exhibition, San Diego, Calif , USA, March 2012.

3.   A. Malekpour and B. W. Karney, "Rapid filling analysis of pipelines with undulating profiles by the method of characteristics," ISRN Applied Mathematics, vol. 2011, Article ID 930460, 16 pages, 2011. ·

4.   H. Santos, C. Leuchtenberg, and S. Shayegi, "Micro-flux control: the next generation in drilling process," in Proceedings of the SPE Latin and Caribbean Petroleum Engineering Conference, Paper SPE 81183, Port-of-Spain, Trinidad, West Indies, 2003.

5.   H. Guner, Simulation study of emerging well control methods for influxes caused by bottomhole pressure fluctuations during managed pressure drilling, Louisiana State University, Baton Rouge, La, USA, 2009.

6.   G. M. D. Oliveira, A. T. Franco, C. O. R. Negrão, A. L. Martins, and R. A. Silva, "Modeling and validation of pressure propagation in

drilling fluids pumped into a closed well," Journal of Petroleum Science and Engineering, vol. 103, pp. 61–71, 2013. ·

7.   I. Zubizarreta, "Pore pressure evolution, core damage and tripping out schedulesf: a computational fluid dynamics approach," in Proceedings of the SPE/IADC Drilling Conference and Exhibition, Amsterdam, The Netherlands, 2013.

8.   J. I. Hage and D. ter Avest, "Borehole acoustics applied to kick detection," Journal of Petroleum Science and Engineering, vol. 12, no. 2, pp. 157–166, 1994. · ·

9.   P. Thodi, M. McQueen, M. Paulin, G. Lanan, and INTECSEA, "Theory and Application of Probabilistic Method of Designing Customized Interval Control Valves Choke Trim for Multizone Intelligent Well Systems," 2007.

10.   A. Mallock, "The damping of sound by frothy liquids," Proceedings of the Royal Society A: Mathematical, Physical and Engineering Sciences, vol. 84, no. 572, pp. 391–395, 1910.

11.   A. B. Wood, A Textbook of Sound, G. Bill and Sons, London, UK, 1941.

12.   E. L. Carstensen and L. L. Foldy, "Propagation of sound through a liquid containing bubbles," The Journal of the Acoustical Society of America, vol. 19, no. 3, pp. 481–501, 1941.

13.   R. A. Thuraisingham, "Sound speed in bubbly water at megahertz frequencies," Ultrasonics, vol. 36, no. 6, pp. 767–773, 1998. · ·

14.   D. Hsieh and M. S. Plesset, "On the propagation of sound in a liquid containing gas bubbles," Physics of Fluids, vol. 4, no. 8, pp. 970–975, 1961. · · ·

15.   J. D. Murray, "Note on the propagation of disturbances in a liquid containing gas bubbles," Applied Scientific Research, vol. 13, no. 1, pp. 281–290, 1964. · ·

16.   G. B. Wallis, One Dimensional Two Phase Flow, vol. 7, McGraw-Hill, New York, NY, USA, 1969.

17.   F. J. Moody, "A pressure pulse model for two-phase critical flow and sonic velocity," Journal of Heat Transfer, vol. 91, no. 3, pp. 371–384, 1969.

18. D. F. D'Arcy, "On acoustic propagation and critical mass flux in two-phase flow," Journal of Heat Transfer, vol. 93, no. 4, pp. 413–421, 1971. ·

19. D. McWilliam and R. K. Duggins, "Speed of sound in a bubbly liquids," Proceedings of the Institution of Mechanical Engineers C, vol. 184, no. 3, pp. 102–107, 1969.

20. R. E Henry, M. A. Grolmes, and H. K. Fauske, "Pressure-pulse propagation in two-phase one- and two-component mixtures," Reactor Analysis and Safety Division, Argonne National Laboratory, 1971.

21. C. S. Martin and M. Padmanabhan, "Pressure pulse propagation in two-component slug flow," Journal of Fluids Engineering, vol. 101, no. 1, pp. 44–52, 1979. · ·

22. Y. Mori, K. Hijikata, and A. Komine, "Propagation of pressure waves in two-phase flow," International Journal of Multiphase Flow, vol. 2, no. 2, pp. 139–152, 1975. · ·

23. Y. Mori, K. Huikata, and T. Ohmori, "Propagation of a pressure wave in two-phase flow with very high void fraction," International Journal of Multiphase Flow, vol. 2, no. 4, pp. 453–464, 1976. · ·

24. D. L. Nguyen, E. R. F. Winter, and M. Greiner, "Sonic velocity in two-phase systems," International Journal of Multiphase Flow, vol. 7, no. 3, pp. 311–320, 1981. · ·

25. R. C. Mecredy and L. J. Hamilton, "The effects of nonequilibrium heat, mass and momentum transfer on two-phase sound speed," International Journal of Heat and Mass Transfer, vol. 15, no. 1, pp. 61–72, 1972. · · ·

26. E. E. Michaelides and K. L. Zissis, "Velocity of sound in two-phase mixtures," International Journal of Heat and Fluid Flow, vol. 4, no. 2, pp. 79–84, 1983. · ·

27. H. W. Zhang, Z. P. Liu, D. Zhang, and Y. H. Wu, "Study on the sound velocity in an aerated flow,"Journal of Hydraulic Engineering, vol. 44, no. 9, pp. 1015–1022, 2013.

28. L. Y. Cheng, D. A. Drew, and R. T. Lahey Jr., "An analysis of wave propagation in bubbly two-component, two-phase flow," Journal of Heat Transfer, vol. 107, no. 2, pp. 402–408, 1985. · ·

29. A. E. Ruggles, R. T. Lahey Jr., D. A. Drew, and H. A. Scarton, "An investigation of the propagation of pressure perturbations in bubbly air/water flows," Journal of Heat Transfer, vol. 110, no. 2, pp. 494–499, 1988. · ·

30. A. E. Ruggles, R. T. Lahey Jr., D. A. Drew, and H. A. Scarton, "The Relationship between standing waves, pressure pulse propagation, and critical flow rate in two-phase mixtures," Journal of Heat Transfer, vol. 111, no. 2, pp. 467–473, 1989. · ·

31. N. M. Chung, W. K. Lin, B. S. Pei, and Y. Y. Hsu, "A model for sound velocity in a two-phase air-water bubbly flow," Nuclear Technology, vol. 99, no. 1, pp. 80–89, 1992.

32. S. Lee, K. Chang, and K. Kim, "Pressure wave speeds from the characteristics of two fluids, two-phase hyperbolic equation system," International Journal of Multiphase Flow, vol. 24, no. 5, pp. 855–866, 1998. · ·

33. J. F. Zhao and W. Li, "Sonic velocity in gas-liquid two-phase flows," Journal of Basic Science and Engineering, vol. 7, no. 3, pp. 321–325, 1999.

34. L. Liu, Y. Wang, and F. Zhou, "Propagation speed of pressure wave in gas-liquid two-phase flow,"Chinese Journal of Applied Mechanics, vol. 16, no. 3, pp. 22–27, 1999.

35. J. L. Xu and T. K. Chen, "Acoustic wave prediction in flowing steam-water two-phase mixture,"International Journal of Heat and Mass Transfer, vol. 43, no. 7, pp. 1079–1088, 2000. · ·

36. C. E. Brennen, Fundamentals of Multiphase Flows, Cambridge University Press, 2005.

37. G. Yeom and K. Chang, "The wave characteristics of two-phase flows predicted by HLL scheme using interfacial friction terms," Numerical Heat Transfer A: Applications, vol. 58, no. 5, pp. 356–384, 2010. · ·

38. Y. Li, C. Li, E. Chen, and Y. Ying, "Pressure wave propagation characteristics in a two-phase flow pipeline for liquid-propellant rocket," Aerospace Science and Technology, vol. 15, no. 6, pp. 453–464, 2011. · ·

39. W. Kuczyński, "Characterization of pressure-wave propagation during the condensation of R404A and R134a refrigerants in pipe mini-channels that undergo periodic hydrodynamic

disturbances,"International Journal of Heat and Fluid Flow, vol. 40, pp. 135–150, 2013. · ·

40. Z. S. Fu, "The gas cut detect technology during drilling in Soviet Union," Petroleum Drilling Techniques, vol. 18, no. 1, pp. 19–21, 1990.

41. X. F. Li, C. X. Guan, and X. X. Sui, "The theory of gas influx detection of pressure wave and its application," Acta Petrolei Sinica, vol. 18, no. 3, pp. 128–133, 1997.

42. X. Wang, "Transmission velocity model in frequency domain of drilling-fluid consecutive pulse signal,"Journal of Southwest Petroleum University, vol. 31, no. 3, pp. 138–141, 2009. · ·

43. M. Davoudi, J. R. Smith, B. Patel, and J. E. Chirinos, "Evaluation of alternative initial responses to kicks taken during managed pressure drilling," in Proceedings of the IADC/SPE Drilling Conference and Exhibition, SPE-128424-MS, New Orleans, La, USA, February 2010. ·

44. H. Li, G. Li, Y. Meng, G. Shu, K. Zhu, and X. Xu, "Attenuation law of MWD pulses in aerated drilling,"Petroleum Exploration and Development, vol. 39, no. 2, pp. 250–255, 2012. · ·

45. L. Xiushan, "Multiphase simulation technique of drilling fluid pulse transmission along well bore," Acta Petrolei Sinica, vol. 27, no. 4, pp. 115–118, 2006.

46. F. Huang, M. Takahashi, and L. Guo, "Pressure wave propagation in air-water bubbly and slug flow,"Progress in Nuclear Energy, vol. 47, no. 1–4, pp. 648–655, 2005. · ·

47. J.-W. Park, D. A. Drew, and R. T. Lahey Jr., "The analysis of void wave propagation in adiabatic monodispersed bubbly two-phase flows using an ensemble-averaged two-fluid model," International Journal of Multiphase Flow, vol. 24, no. 7, pp. 1205–1244, 1998. · ·

48. G. B. Wallis, One Dimensional Two Phase Flow, vol. 7, McGraw Hill, New York, NY, USA, 1969.

49. L. Yarborough and K. R. Hall, "How to solve equation of state for z-factors," Oil and Gas Journal, vol. 72, no. 7, pp. 86–88, 1974.

50. A. R. Hasan and C. S. Kabir, "Wellbore heat-transfer modeling and applications," Journal of Petroleum Science and Engineering, vol. 86–87, pp. 127–136, 2012. · ·

51. M. Al-Oufi Fahd, An investigation of gas void fraction and transition conditions for two-phase flow in an annular gap bubble column [Ph.D. thesis], Loughborough University, 2011.

52. M. J. Sanchez, Comparison of correlations for predicting pressure losses in vertical multiphase annular flow [M.S. thesis], The University of Tulsa, 1972.

53. Y. Taitel, D. Bornea, and A. E. Dukler, "Modelling flow pattern transitions for steady upward gas-liquid flow in vertical tubes," AIChE Journal, vol. 26, no. 3, pp. 345–354, 1980. · ·

54. G. Costigan and P. B. Whalley, "Slug flow regime identification from dynamic void fraction measurements in vertical air-water flows," International Journal of Multiphase Flow, vol. 23, no. 2, pp. 263–282, 1997. · · ·

55. V. Casariego and A. T. Bourgoyne, "Generation, migration, and transportation of gas-contaminated regions of drilling fluid," in Proceedings of the SPE Annual Technical Conference and Exhibition, Houston, Tex, USA, October 1988.

56. T. Z. Harmathy, "Velocity of large drops and bubbles in media of re-stricted or infinite extent," AIChE Journal, vol. 6, no. 2, pp. 281–288, 1960.

57. A. R. Hasan and C. S. Kabir, "Predicting multiphase flow behavior in a deviated well," SPE Production Engineering, vol. 3, no. 4, pp. 474–482, 1988. · ·

58. X. W. Kong, Y. H. Lin, and Y. J. Qiu, "A new method for predicting the position of gas influx based on PRP in drilling operations," Journal of Applied Mathematics, vol. 2014, Article ID 969465, 12 pages, 2014. ·

Simulation of CO_2 and H_2S Removal Using Methanol in Hollow Fiber Membrane Gas Absorber (HFMGA)

Majid Mahdavian[1], Hossein Atashi[1],
Morteza Zivdar[1], and Mahmood Mousavi[2]

[1]Department of Chemical Engineering, University of Sistan and Baluchestan, Zahedan, Iran

[2]Department of Chemical Engineering, Ferdowsi University of Mashhad, Mashhad, Iran

ABSTRACT

Application of methanol solvent for physical absorption of CO_2 and H_2S from $CO_2/H_2S/CH_4$ mixture in gas-liquid hollow fiber membrane gas absorber (HFMGA) was investigated. A computational mass transfer (CMT) model for simulation of HFMGA in the case of simultaneous separation of CO_2 and H_2S was developed. The membrane gas absorber model explicitly calculates for the rates of mass transfer through the membrane and components concentration profiles. Due to the lack of experimental data in the literature, the model was validated using available individual components' water absorption data. The numerical predictions were in good agreement with the experimental data. The effects of operating conditions such as liquid velocity, gas velocity, temperature and pressure were analyzed. It is shown that methanol solvent can successfully be used for CO_2 and H_2S removal in membrane gas absorber. Also it is found that the concentration distribution of CO_2 and H_2S in the gas phase along the fiber length obeys plug flow model whereas in the methanol absorbent deeply affected by the interface concentration, absorbent velocity and diffusivity. In addition, it is shown that application of membrane gas absorber using methanol absorbents for H_2S removal and at higher flow rate is more efficient. Moreover, at operating pressures above 10 atm even at low absorbent rate, H_2S concentration depletion is relatively complete while at 1 atm this value is about 30%. This means that removal efficiency decreases with an increase in temperature and it is more important especially for H_2S.

INTRODUCTION

Some industrial gas streams (such as natural gas processing, petroleum refineries, petrochemicals) frequently contain H_2S and CO_2 as impurities. All of these gases requires treatment before delivery to the pipeline. It is reported that CO_2 is representing about 80% of greenhouse gases and half of the CO_2 emissions are produced by industrial plants such as fossil-fuel-fired power plants, iron, steel and cement works [1]. Also carbon dioxide is a common contaminant of natural gas and must be removed to a level of <8% (usually <2%) to minimize corrosion of the pipeline. Hydrogen sulfide removal is also desirable to reduce corrosion. In many cases it is necessary from the health and safety standpoint [2].

The most well known technology for recovery/removal of CO_2 and H_2S is solvent absorption. This technology was established over 80 years ago in the chemical and oil industrials for the removal of acid gases from natural gas streams. For the removal of CO_2 and H_2S, traditionally absorption processes like packed and plate columns are utilized [3]. Because these generally require large space and high investment cost, the emphasis of designing most of these operations is towards maximizing the mass transfer rate by creating as much interfacial area as possible [4]. In addition, they also suffer from several limitations including flooding, loading entrainment, foaming, weeping, etc. In recent years, the demand for alternative technologies has increased and many researchers have looked for new technologies to enhance the efficiency of absorption processes. Membrane-based absorption technique has been introduced as an emerging technology for the recovery/removal of gases (like CO_2, H_2S, SO_2, NH_3, VOC, etc.) from various industrial process gas streams [5]. In addition to gas/liquid, this technology also has found applications in numerous liquid/liquid applications such as fermentation, pharmaceuticals, wastewater treatment, semiconductor manufacturing, carbonation of beverages, metal ion extraction, protein extraction, osmotic distillation and other operations [6].

Membrane gas absorbers are devices that achieve two phase mass transfer through diffusion without dispersing one phase within another. Such a device employs a porous membrane acts as a non-selective barrier between both phases where the gas and the absorbent solution flow on two sides of a membrane [5]. The membranes are usually microporous and can be both hydrophobic and hydrophilic. Hydrophobic microporous membranes like polypropylene (PP), polytetrafluoroethylene (PTFE) and polyvinylidene fluoride (PVDF) membranes have received increasing attention in recent years for using in membrane gas absorbers because of their good hydrophobicity [7]. These membrane absorber systems, generally in the form of hollow fibers with diameters of 0.5 mm - 1 mm in densely packed membrane modules, provide a high interfacial area (500 m^2/m^3 - 2000 m^2/m^3) significantly greater than most traditional absorbers (100 m^2/m^3 - 800 m^2/m^3) between two phases to achieve high overall rates of mass transfer. This significantly decreases the size required for the contactor [8]. Moreover, this kinds of contacting devices offers a number of important advantages over conventional dispersed phase contactor for gas sorption, such as large interfacial area between gas and liquid flow (up to two orders of magnitude more surface area per volume than conventional contactors), no flooding and foaming phenomena, independent control of gas and liquid flow rates, high efficiency, the possibility of combining absorption and desorption in one single compact module, energy intensive, and so on (as an example of review, see Gabelman et al. [5]).

Chemical absorbents like amines and amino acid salts are extensively used in the removal of impurities from gas mixtures. Physical absorbents have been of considerable interest in the development of gas treatment solvents, especially when the partial pressure of undesirable impurity is high. Some of the physical solvents used commercially are propylene carbonate (PC), n-formyl morpholine (NFM), dimethyl ethers of polyethylene glycol (DEPG), and n-methyl-pyrrolidone (NMP) (see more example in [9]). Physical solvents can be a possible alternative to chemical solvents in certain areas of applications, although they are less effective than chemical absorbents (i.e. the specific absorption rate into physical absorbents in comparison with chemical solvents is less). But they can be regenerated by just pressure reduction method without large amount of heat supply and thus excessive energy savings can be obtained [9]. An economical analysis must be done to select the best choice of solvent. In addition, they can be used as pre-treatment solvent in the development of hybrid systems. The most well known physical absorbent is water. However, its economics are limited by the relatively low solubility which leads to larger amounts of circulation rate, i.e. the higher investment costs as well as the higher operating costs [10]. However, there are good organic solvents which possess a much higher solvent capacity than water. Among the physical solvents, n-methyl-2-pyrrolidone (NMP), methanol and propylene carbonate (PC) are popular as gas treating solvents. Methanol has a high thermal and chemical stability, low vapor pressure, and is not corrosive. It is able to absorb acid gases, hydrocarbons, mercaptans and water. Moreover, it is produced in big quantity and readily available [10]. This properties make it highly effective for processing a wide range of compositions.

The applications of hollow fiber gas-liquid membrane gas absorber for acid gas removal specially carbon dioxide from gas mixtures have been studied by several researchers. In this case, a large number of experimental absorption studies and theoretical modeling analyses have been performed with physical or chemical absorbent liquids such as pure water, aqueous amine solution, aqueous sodium hydroxide solution, aqueous potassium carbonate solution, aqueous blended solvents, etc. [11- 14]. Some authors have explored possible simultaneous removal of H_2S and CO_2 in hollow fiber membrane gas absorber using MEA [1] and DEA [15]. However, of the authors considering chemical absorption, few have worked with physical solvents as would be the good choice in membrane gas absorber process. There have been few attempts to address possible physical absorption in hollow fiber membrane gas absorbers [16-18] that mostly describes the water performance and theoretical analysis of simultaneous removal of CO_2 and H_2S using methanol absorbent in HFMGA has not been discussed by researchers.

In the present work, after modification of 2D mathematical model, this new process has been applied for CO_2 and H_2S capture from carbon dioxide/hydrogen sulfide/ methane mixture (when the partial pressures of CO_2 and H_2S are 10% of total pressure) using methanol (as an example of physical absorbent) absorbent and its potential possibility for carbon dioxide and hydrogen sulfide removal has been evaluated. It should be mentioned that areas of possible HFMGA process for gas treatment using physical solvent with economic considerations will be reported in another work. This work was performed using CFD tool with respect to solubility behavior. CFD has been largely used as a powerful tool to model membrane separation processes. It is able to simulate the concentration, temperature and velocity fields as well as the transport parameters and operating efficiency.

MODEL DEVELOPMENT

In this paper, a steady-state two-dimensional mathematical model has been modified (e.g. [1,12]) to describe the physical absorption of carbon dioxide and hydrogen sulfide in the polymeric hollow fiber membrane gas absorber (using methanol absorbent as the absorption liquid). The model describes the mass transfer in the gas, membrane and liquid phases. Axial and radial diffusion inside the shell, through the membrane, and within the tube side of the membrane gas absorber have been considered in the model equations. It allows studying the effect of membrane wetting on the mass transfer through the membrane and also the effect of operating conditions (gas and liquid flow rates, temperature), solvent affinity (H) and flow pattern (counter current or co current arrangement) on the carbon dioxide and hydrogen sulfide removal efficiencies.

This model assumes that the fibers are distributed evenly through the shell space, which allows the results obtained with a single fiber to be generalized to the entire module. Model results are based on "non-wetted mode" in which the gas mixture filled the membrane pores.

The following assumptions are made to develop the governing mass transfer differential equations: 1) fully developed parabolic velocity profile in the hollow fiber under laminar flow conditions; 2) the mixture gases flow inside the shell are ideal gas; 3) Happel's free surface model [19] is used to characterize the velocity profile at the shell side; 4) the physical properties of the fluid are constant; 5) the Henry's law is applicable for gas-liquid interface; 6) no absorption of bulk and inert gases; 7) pitch and placing of the fibers are uniform; 8) no pore blockage.

Transport Model for the Hollow Fiber Membrane Gas Absorber

In order to describe the mass transfer and develop the equations of mathematical model in the hollow fiber membrane gas absorber, a material balance has been applied for a segment of a hollow fiber, as shown in Figure 1 in the shell, membrane and tube sides. Also, the computational domain used for the numerical simulation is shown in Figure 1. This model is based on the idea that two concentric cylinders are used as the model for fluid flowing out of the fibers and so only portion of fluid surrounding the fiber is considered and may be approximated as circular cross section [19] The fluid flow is described using the fully developed laminar flow model in the tube side, whereas the fluid flow in the shell side is characterized by the Happel's free surface model.

The position $r = 0$ is the center of the fiber and r_1, r_2 and r_3 are the inner, outer and Happel's free model radii of the fiber, respectively (Figure 1). The radius of Happel's free surface model is calculated to be $r_3 = 720$ μm. Dimensions of the hollow-fiber membrane gas absorber are listed in Table 1. The gas mixture consists of carbon dioxide, hydrogen sulfide and methane is fed to the shell side at $z = L$, while the liquid (methanol) is passed through the tube side at $z = 0$. Carbon dioxide and hydrogen sulfide are removed from the mixture by diffusing through the membrane due to a concentration gradient and then absorbing with the solvent.

Equations Describing the Shell Side

Convective-diffusion equation for the component i using Fick's law of diffusion, when chemical reaction is taking place, can be written as:

$$\partial C_{i,shell} / \partial t + \nabla \cdot \left(-D_{i,shell} \nabla C_{i,shell} - C_{i,shell} V_{i,shell} \right) = R_{i,shell} \tag{1}$$

where C_i, $D_{i,shell}$, R_i and V_{shell} denote the local concentration of the component i, the diffusivity of the component i, reaction rate of the component i and axial velocity in shell side, respectively. According to the Happel's free surface model [19], the velocity profile in the shell side may be obtained. For this purpose, a momentum balance over a thin cylindrical shell is integrated twice to obtaining the following equation for the shell side velocity distribution, which have been applied by several authors (e.g. [20, 21]):

$$V_{shell} = 2 V_{ave-shell} \left(1 - \left(\frac{r_2}{r_3} \right)^2 \right)$$

$$\cdot \left(\frac{\left(r/r_3 \right)^2 - \left(r_2/r_3 \right)^2 + 2 \ln \left(r_2/r \right)}{3 + \left(r_2/r_3 \right)^4 - 4 \left(r_2/r_3 \right)^2 + 4 \ln \left(r_2/r_3 \right)} \right) \tag{2}$$

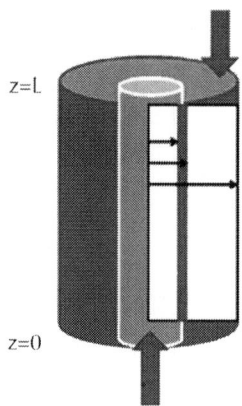

Figure 1: The schematic diagram of a hollow fiber membrane and computational domain.

Table 1: Specifications of the membrane gas absorber

Parameter	Value
Module I.D. (m)	4.35×10^{-3}
Module O.D. (m)	6.35×10^{-3}
Fiber O.D. (m)	9.07×10^{-4}
Fiber I.D. (m)	6.07×10^{-4}
Module length (m)	0.3
Fiber length (m)	0.2725
No. of fibers	9

where V_{shell} is the velocity component inside the shell, $V_{ave-shell}$ is the shell average velocity in the axial direction, r_2(m) is the outer fiber radius and r_3(m) is Happel's free surface model radius defined as:

$$r_3 = r_2 \sqrt{1/(1-\phi)}$$

(3)

Packing density (ϕ) can be defined as the ratio of total surface area of membrane to the cross-sectional area of the module and ϕ is calculated as:

$$\phi = n \left(\frac{r_2}{r_4} \right)^2$$

(4)

where n is the number of fibers and r_4 is the inner radius of the MGA. The partial differential equation of the steady state mass balance for cylindrical coordinates, where no reaction takes place in the shell side is obtained and it is given as follows:

$$D_{i-shell} \left(\frac{\partial^2 C_{i-shell}}{\partial r^2} + \frac{1}{r} \frac{\partial C_{i-shell}}{\partial r} + \frac{\partial^2 C_{i-shell}}{\partial z^2} \right) = V_{shell} \frac{\partial C_{i-shell}}{\partial z}$$

(5)

The boundary conditions are the following ($i = CO_2$, H_2S)

$$\text{At } z = L : C_{i-shell} = C_{0,i-shell}$$

(8)

$$At \ r = r_2 : C_{i-\text{shell}} = C_{i-\text{mem}}$$

(7)

$$At \ r = r_3 : \partial C_{i-\text{shell}} / \partial r = 0 \ (\text{symmetry})$$

(8)

Equations Describing the Membrane Side

Mass transfer takes place through the membrane pores without mixing between phases and the transfer equation inside the pores can be derived without considering convection. The membrane diffusivity of species within the pores should be employed instead of the ordinary diffusivity. This parameter can be defined as $D_{i-\text{mem}} = D_i \varepsilon / \tau$. The steady state material balance for the transport of diffusing components (i = CO_2, H_2S) inside the membrane can be considered for non-wetting condition, where pores filled by the gas phase. For the non-wetting case, no chemical reactions will be considered in the membrane.

$$D_{i-\text{mem}} \left(\frac{\partial^2 C_{i-\text{mem}}}{\partial r^2} + \frac{1}{r} \frac{\partial C_{i-\text{mem}}}{\partial r} + \frac{\partial^2 C_{i-\text{mem}}}{\partial z^2} \right) = 0$$

(9)

$$At \ r = r_1 : C_{i-\text{mem}} = C_{i-\text{tube}} / m_i$$

(10)

$$At \ r = r_2 : C_{i-\text{mem}} = C_{i-\text{shell}}$$

(11)

In the partial wetted mode, additional equations are required to describe diffusion-reaction inside the wetted parts of the pores. In this case the pores are both gas and liquid-filled (wetted and non-wetted parts) and the transport of the species i generally depends on its diffusion coefficient into the liquid [22].

Non-Wetted Part of the Membrane

$$D_{i-drymem}\left(\frac{\partial^2 C_{i-drymem}}{\partial r^2}+\frac{1}{r}\frac{\partial C_{i-drymem}}{\partial r}+\frac{\partial^2 C_{i-drymem}}{\partial z^2}\right)=0$$

(12)

Boundary conditions are:

$$At\ r=r_2 : C_{i-drymem}=C_{i-shell}$$

(13)

$$At\ r=r_w : C_{i-drymem}=C_{i-wetmem}/m_i$$

(14)

Wetted Part of the Membrane

$$D_{i-wetmem}\left(\frac{\partial^2 C_{i-wetmem}}{\partial r^2}+\frac{1}{r}\frac{\partial C_{i-wetmem}}{\partial r}+\frac{\partial^2 C_{i-wetmem}}{\partial z^2}\right)+\varepsilon R_i=0$$

(15)

Boundary conditions are:

$$At\ r=r_w : C_{i-wetmem}=m_i C_{i-drymem}$$

(16)

$$At\ r=r_1 : C_{i-wetmem}=C_{i-tube}, C_{solvent-tube}=C_{solvent-mem}$$

(17)

where m is dimensionless distribution coefficient.

Equations Describing the Tube Side

The partial differential equation of the steady state mass balance for each species during simultaneous mass transfer in a non-reactive absorption system is obtained and can be expressed as:

$$D_{i-tube}\left(\frac{\partial^2 C_{i-tube}}{\partial r^2}+\frac{1}{r}\frac{\partial C_{i-tube}}{\partial r}+\frac{\partial^2 C_{i-tube}}{\partial z^2}\right)=V_{tube}\frac{\partial C_{i-tube}}{\partial z}$$

(18)

The left-hand side of the above equation represents the diffusion and reaction terms, whereas the right-hand side is the convection term.

Considering laminar flow, Navier-Stokes equations and the equation of continuity can be solved for fluid flow in a cylindrical pipe, therefore velocity distribution in the tube in the z direction can then be obtained as [22]:

$$V_{tube} = 2V_{ave-tube}\left(1-\left(\frac{r}{r_1}\right)^2\right)$$

(19)

where V, rand r1 are the average velocity in the lumen, the radial distance and the radius of the lumen, respectively. The following boundary conditions are considered:

$$At \ z = 0 : C_{solvent-shell} = C_{0.solvent-tube}$$

(20)

$$At \ r = 0 : \partial C_{i-tube}/\partial r = 0 \ (symmetry)$$

(21)

$$At \ r = r_1 : C_{i-tube} = m_i C_{i-mem}$$

(22)

Physical Properties and Numerical Solution

Simulation of membrane gas absorber requires data on physicochemical properties used as input parameters in the model such as solubility and diffusivity of the relevant components in each phase. The distribution coefficient of CO$_2$ was taken from Versteeg et al. [23] and distribution coefficient of H$_2$S was taken from Carroll et al. [24] as a function of temperature in water. Henry's constant of CO$_2$ and H$_2$S for methanol as a function of temperature was reported by Lunsford et al. [25]. Liquidphase diffusivities of CO$_2$ [23] and H$_2$S [26] in water were estimated by the equations proposed by Versteeg and Cussler respectively and their value in methanol were estimated using the correlation given by Diaz et al. [27]. Gas-phase diffusivities of CO$_2$ and H$_2$S were estimated using the correlation given by Diaz et al. [27] and Cussler et al. [26]. Gas-filled membrane phase diffusivities were corrected for membrane porosity and tortuosity. The values for other data were obtained from [28, 29].

In order to solve the coupled partial differential equations for the tube, membrane and shell sides with the appropriate boundary conditions and physical and chemical properties for CO_2 and H_2S, FEMLAB software has been used.

RESULTS AND DISCUSSION

Model Validation

The model of simultaneous absorption of CO_2 and H_2S using methanol in hollow fiber membrane gas absorber was validated using available individual components' absorption data of Al-Marzouqi et al. [20] and Faiz et al. [30] for physical absorption of CO_2 and H_2S in water, respectively, since we didn't find any reported experimenttal data for the type of the present work in the literature.

Comparison of the experimental data and the simulated results are shown in Figures 2 and 3. It should be noted that the type of hollow fiber MGA modules and the operating conditions applied for obtaining the mentioned experimental data differ significantly with each other and we applied the exact conditions for each case. Results of both validations for physical absorption of CO_2 and H_2S are shown in Figures 2 and 3, respectively. Figure 2 shows the calculation of the model with the experimental results of percent CO_2 removal as a function of water flow rate at the gas flow rate of 200 ml/min and Figure 3 shows the percent H_2S removal and the outlet H_2S concentration as a function of inlet concentration of H_2S in the gas phase at gas velocity of 5.1 m/s and liquid velocity of 0.092 m/s. Clearly the model predictions are in good agreement with the two set of CO_2 and H_2S absorption data and it shows that the numerical model accurately predicts the experimental data for both gases.

Concentration Distribution of CO_2 and H_2S

A steady state component concentration distribution is established inside the shell, membrane, and within the tube side of the hollow fiber membrane gas absorber that affects mass transfer coefficient, removal efficiency and mass transport. Numerically calculated carbon dioxide and hydrogen sulfide concentration distribution in each of the three phases are depicted in Figures 4 and 5, respectively. In order to compare the difference between concentration profiles of CO$_2$ and H$_2$S effectively, equal input concentration in gas phase is applied. The solubility of CO$_2$ and H$_2$S in methanol is linearly proportional to their partial pressure in the gas mixture and, hence, it can be modeled according to Henry's law. As expected, it can be seen that the concentration near the membrane-liquid wall signifycantly affected by the interface concentration, whereas the CO$_2$ and H$_2$S concentrations on the shell side slightly decrease in the radial direction. The concentration profile is discontinuous at the gas filled membrane-liquid interface based on the equilibrium relationship.

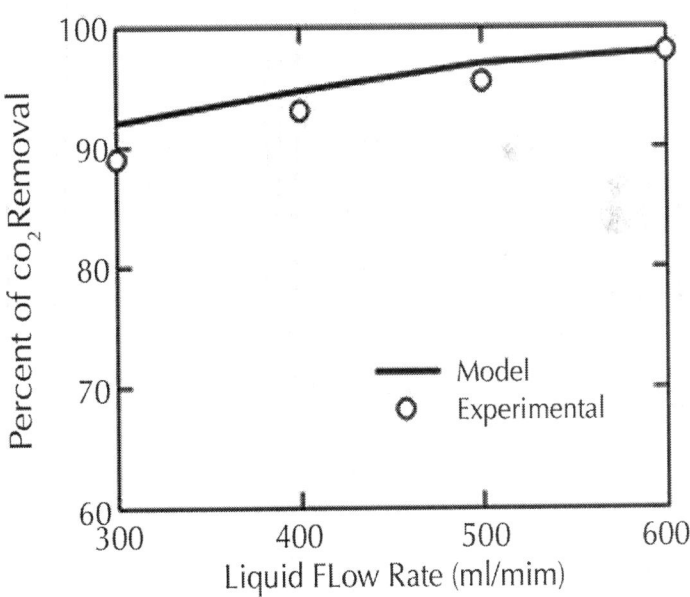

Figure 2: Comparison between experimental [20] and simulated CO$_2$ removal efficiency.

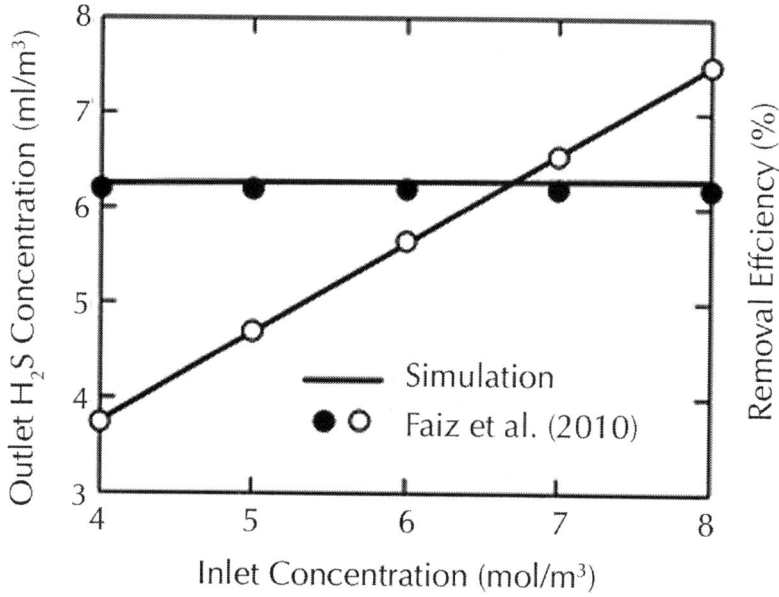

Figure 3: Comparison between model data [30] and simulated H$_2$S gas outlet concentration and removal efficiency.

Due to the dimensions of the hollow fiber, the computational domain is the area of membrane length multiplied by Happel's free surface model width. It is important to note that since the fiber is 900 times longer than its radial dimension (in this case, 0.3 mm in radius and 27 cm in length), a scaling factor of 90 has been applied in the z direction in order to reduce computational cost.

It is worth mentioning that the sensitivity grid-dependence analysis of the method of solution to the mesh size was performed in order to ensure that the numerical solution is not affected by the specification of the mesh size.

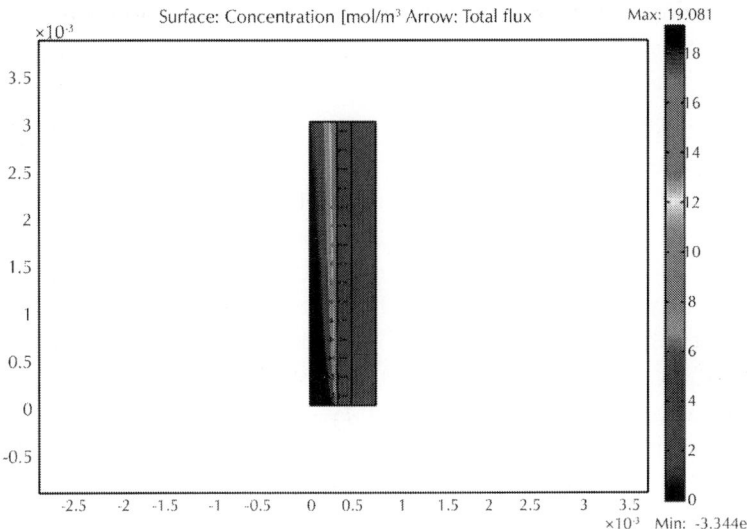

Figure 4: CO$_2$ concentration distribution in computational domain for V_L = 0.1 m/s, V_G = 3 m/s and T = 298 K.

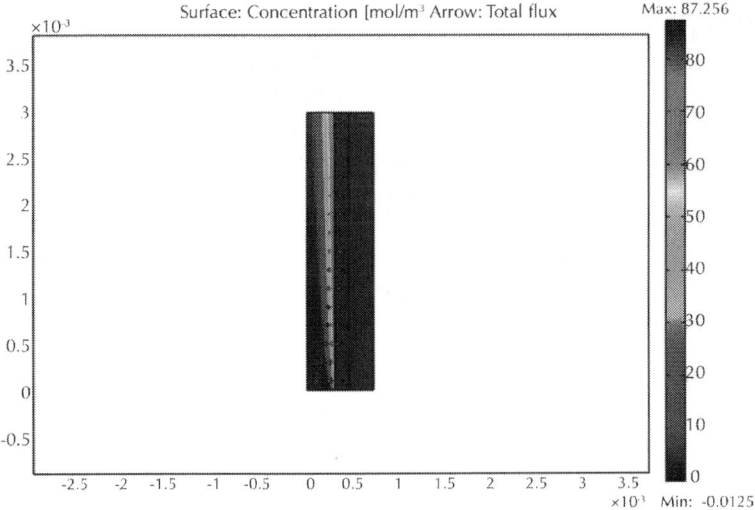

Figure 5: H$_2$S concentration distribution in computational domain for V_L = 0.1 m/s, V_G = 3 m/s and T = 298 K.

The importance of a fine mesh adjacent to the membrane wall is obvious from the components concentration distribution shown in Figures 4 and 5.

Figures 6 and 7 show the numerically calculated dimensionless radial concentration profile of CO_2 and H_2S as a function of dimensionless length at different cross sections along fiber length, i.e. z/L = 0.1, 0.5, 0.9 (Figure 6) and at two different absorbent velocities (Figure 7). With respect to these figures, there is a concentration drop close to the absorbent-membrane interface at the membrane wall in the methanol absorbent phase for both gases. Concentration depletion for CO_2 and H_2S in liquid phase has the same trend but there is a sharper reduction in H_2S concentration in comparison with CO_2 concentration which is attributed to higher solubility of H_2S in the methanol absorbent.

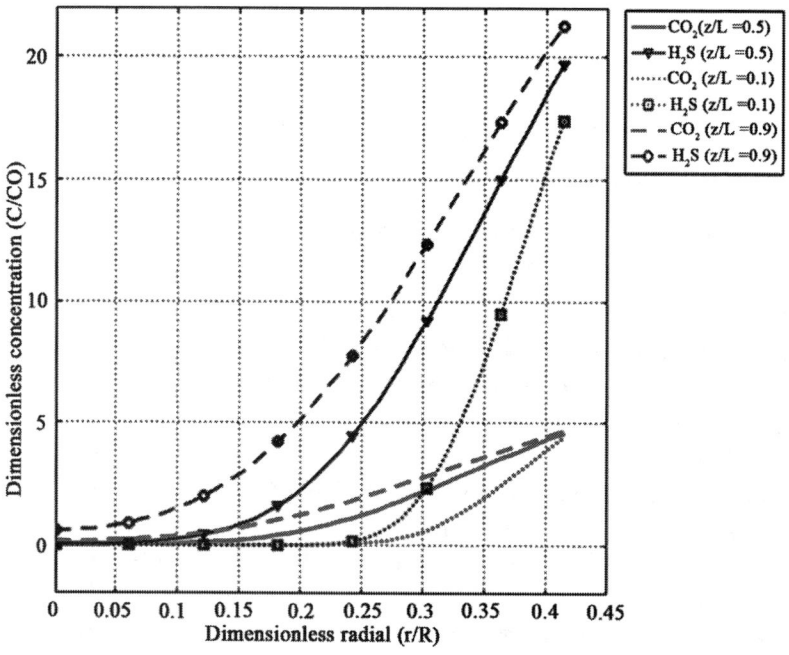

Figure 6: CO_2 and H_2S radial tube side concentration profiles along fiber length for z/L = 0.1, 0.5, 0.9, V_L = 0.1 m/s, V_G = 3 m/s and T = 298 K.

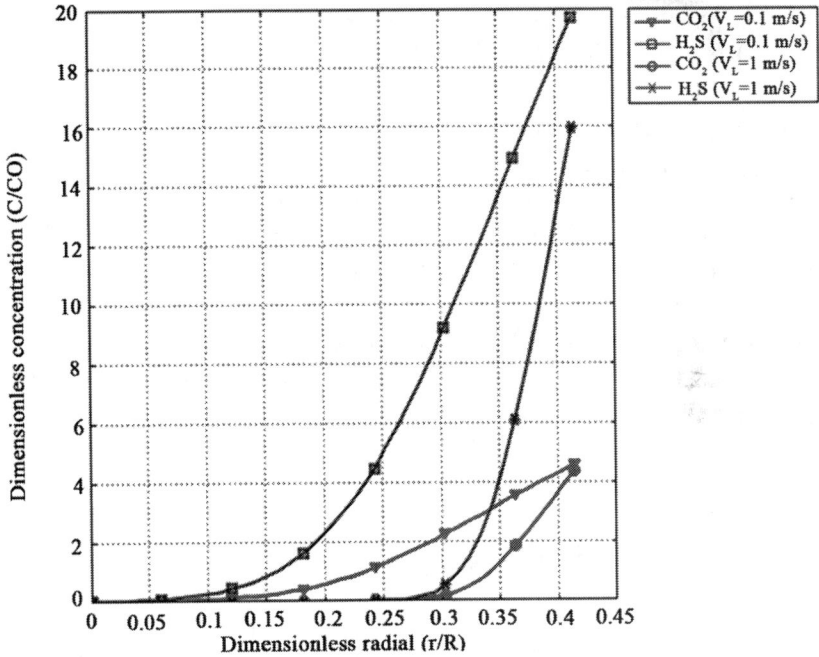

Figure 7: Effect of absorbent velocity on CO_2 and H_2S radial tube side concentration profiles for $V_G = 3$ m/s, z/L = 0.5 and T = 298 K.

The results indicate that penetration depth increases with distance from methanol absorbent entrance (z/L = 0) and, therefore, components diffuse into liquid phase from membrane interface. Note that since liquid phase is very thin, it acts as film layer. With respect to the components diffusivities, liquid velocity and dimension of fiber, it is seen that the contact time is not enough that diffusion entirely affects the liquid phase and absorbed species do not distribute rapidly in radial direction before absorbent leaves the fiber (dimensionless Gz number conception). However, at higher inlet liquid velocity, the depletion of component concentration is faster. The reason is that the axial convective flow decreases with radial diffusion.

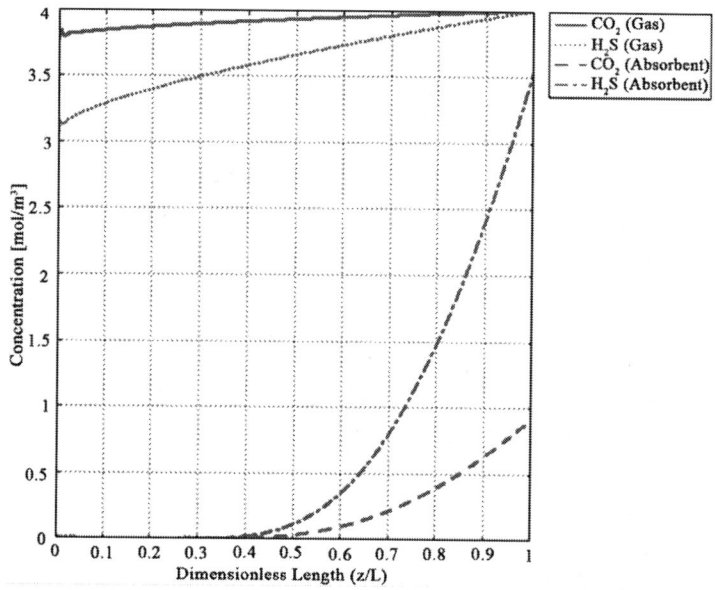

Figure 8: CO_2 and H_2S concentration profile in the axial direction in both shell and tube sides for V_L = 0.1 m/s, V_G = 3 m/s, $C_{0,CO_2=.}$ = C_{0,H_2S} = 4 mol/m^3 and T = 298 K.

Figure 8 shows the axial CO_2 and H_2S concentration profiles in absorbent and gas phase. For absorbent phase, tube center line (r = 0) and for gas phase Happel's radius (r = r_3) is selected. It can be seen that in the case of methanol absorbent, the CO_2 concentration depletion in the gas and amount of absorbed are low in comparison with the H_2S. It obviously indicates the higher capacity of methanol in absorption of H_2S. Based on the bulk concentration of CO_2 and H_2S, removal efficiencies are about 12.9% and 29.3% throughout the fiber, respecttively. Moreover, in 50% and 40% of the fiber length, the concentrations are still zero for CO_2 and H_2S, respectively.

Effect of Absorbent and Gas Velocity on CO_2 and H_2S Removal Efficiencies

Figures 9 and 10 indicate the effect of absorbent and gas velocity on the removal efficiencies of CO_2 and H_2S using methanol absorbent in comparison with water absorbent. Wide range of velocities was selected for both absorbent and gas in order to provide a chance to gain a real insight into this effect. Considering these figures, CO_2 and H_2S removal efficiencies at given conditions increase with the increase in absorbent velocity. This effect is due to the increasing in driving force with entering fresh absorbent. Therefore, CO_2 and H_2S concentrations in gas phase reduce and removal efficiencies improved because of higher absorption rate. This effect is reported by several authors [1, 30, and 31] for water absorbent in the case of physical absorbent in hollow fiber membrane gas absorber devices.

The results show that with increasing the liquid velocity, the overall mass transfer coefficient increases. The reason is that in the case of physical absorption in membrane gas absorber, the controlling resistance for the mass transfer usually is liquid phase. The CO_2 and H_2S removal ability of methanol is illustrated in Figure 9 where the results are plotted for methanol in comparison with water absorbent. for example at the absorbent velocity of 3 m/s, the removal efficiencies using methanol absorbent are 34.7% and 84.3% and removal efficiencies using water absorbent are 13.8% and 21.6% for CO_2 and H_2S, respectively.

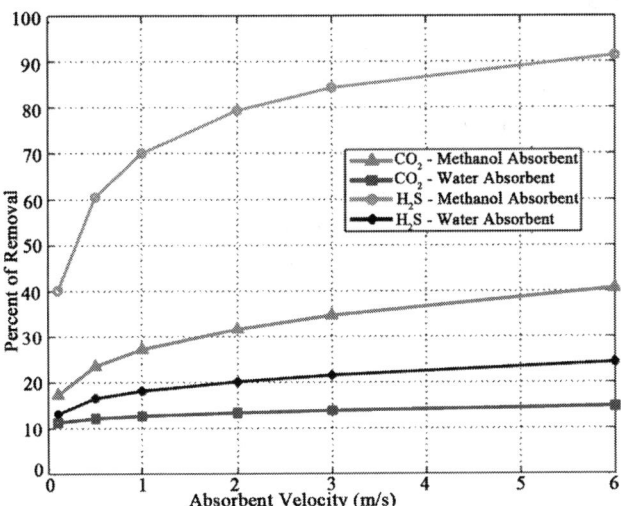

Figure 9: Effect of absorbent velocity on CO_2 and H_2S removal efficiencies for methanol and water absorbent at V_G = 2 m/s, $C_{0,CO_2=.}$ = C_{0,H_2S} = 4 mol/m³ and T = 298 K.

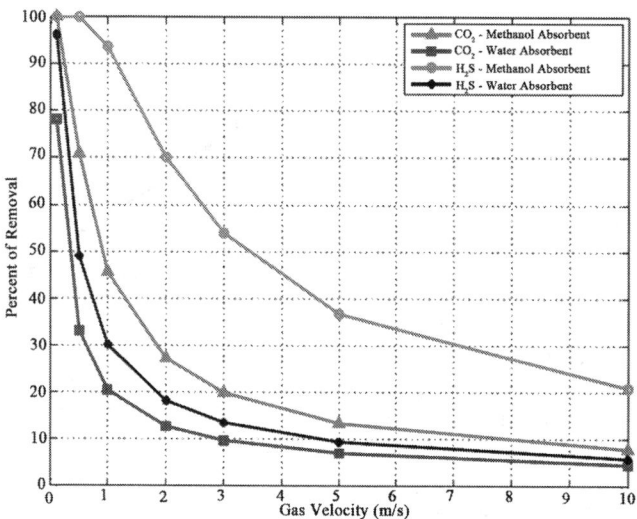

Figure 10: Effect of gas velocity on the CO_2 and H_2S removal efficiencies for methanol and water absorbent at $V_L = 1$ m/s, $C_{0,CO_2} =$. $C_{0,H_2S} = 4$ mol/m^3 and T = 298 K.

It is important to note that in the case of methanol (or water), the CO_2 removal efficiency reaches a relatively constant value whereas, H_2S removal efficiency increases by increasing the liquid velocity which leads to higher relative absorption rates in comparison with CO_2. This is due to the fact that for higher absorbent velocities due to the lower contact time, the absorbent liquid cannot reach saturation and maybe leaves the module unsaturated. In spite of reducing contact time at higher velocities, methanol absorbent (or water) leaves the module saturated with respect to its low CO_2 potential absorption whereas the H_2S potential absorption is high enough resulting in an unsaturated absorbent at the module exit. Relative absorption rate of CO_2 using methanol absorbent is in the range of 1.5 to 2.7 and relative absorption rate of H_2S using methanol absorbent is in the range of 2.9 to 3.75 in comparison with the case of water absorbent in this operating. Therefore, application of membrane gas absorber using methanol absorbent for H_2S removal and at higher flow rate is more efficient. In addition, methanol in comparison with other commercially available physical solvents has a lower viscosity, which increases mass transfer rates and decreases membrane area requirements and pressure drop over the fiber length. Note that in simultaneous absorption of CO_2 and H_2S using methanol in MGA when selective absorption of CO_2 and H_2S is desired, selectivity remains relatively constant with increasing the methanol absorbent flow rate.

Figure 10 shows the effect of gas velocity on the CO_2 and H_2S removal efficiencies for methanol absorbent at a given conditions in comparison with water absorbent. It can be seen that CO_2 and H_2S removal efficiencies decrease considerably with the increase in gas velocity. This effect is due to the fact that by increasing the gas velocity (or flow rate) the amount of input impurity (CO_2 and H_2S) increases at constant absorption ability and on the other hand, gas-liquid contact time decreases. As a result of these two negative effects, CO_2 removal efficiency decreases in the membrane gas absorber.

Note that reduction in removal efficiencies are not the same for equal velocity step size in both gas and absorbent due to the different gas-liquid contact time. For example, contact time decreases 67% when velocity changes from 1 m/s to 3 m/s while it decreases 40% when velocity changes from 3 m/s to 5 m/s.

The Effect of Temperature and Pressure on CO_2 and H_2S Removal Efficiency

Figure 11 shows gas phase CO_2 and H_2S concentration profiles in the axial direction at three different temperatures, i.e. 288 K, 298 K and 308 K. It can be seen that the outlet CO_2 and H_2S concentrations increase and the trend of concentration variations for CO_2 and H_2S are the same but for H_2S is more important: the higher the temperature, the higher the average component concentration in the gas phase and outlet stream (lower removal efficiency). The reason is a result of two opposite effects that as the temperature increases, the solubility of CO_2 and H_2S decrease and liquid-phase diffusion coefficients increase. In addition, temperature effects on CO_2 and H_2S concentration distribution in the radial direction are more important near the membrane-liquid interface in the liquid phase.

Generally, physical solvents are used for undesirable component removal from high-pressure gas streams. Figure 12 shows the effect of pressure on CO_2 and H_2S removal efficiencies for methanol absorbent. In the case of application of membrane gas absorber at high pressures, methanol as a physical solvent is more efficient and it might be an alternative to chemical solvents. At the module exit, complete H_2S concentration depletion is relatively above 10 atm while at 1 atm this value is about 30%. High partial pressure of CO_2 and H_2S or application of physical solvent with high absorption power leads to lower amounts of circulation. For example, the circulation rate need to absorption of CO_2 at a feed pressure of 10 atm in methanol is only about one-fourth of that circulation rate under 1 atm operating pressure.

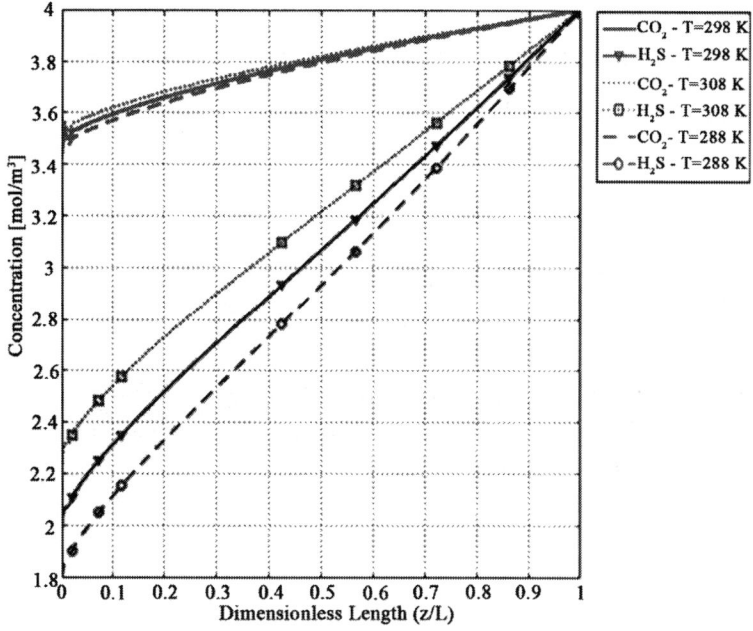

Figure 11: Effect of temperature on CO_2 and H_2S shell side concentration profile for $V_L = 1$ m/s, $V_G = 3$ m/s, $C_{0,CO_2} = C_{0,H_2S} = 4$ mol/m³ and $T = 298$ K.

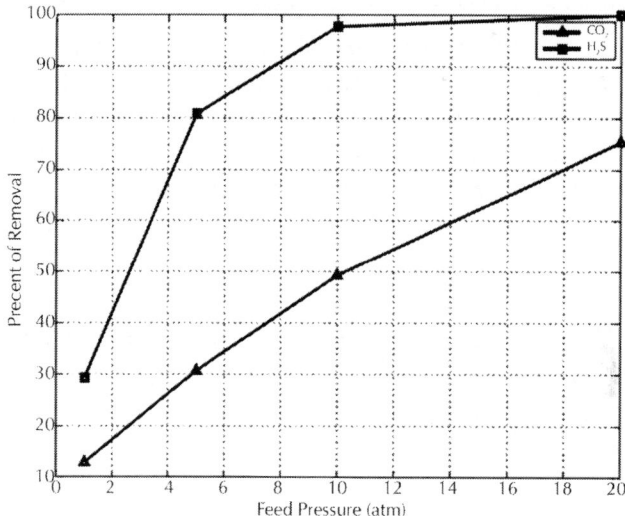

Figure 12: Effect of pressure on CO_2 and H_2S removal efficiencies for Methanol absorbent at $V_L = 0.1$ m/s, $V_G = 3$ m/s and T = 298 K.

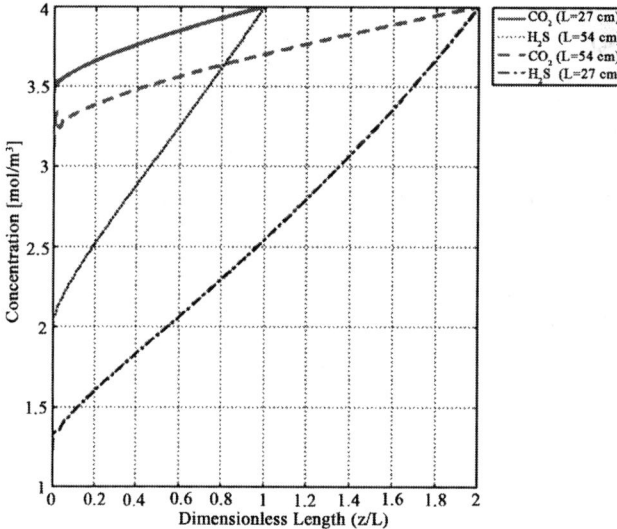

Figure 13: Effect of fiber length on CO_2 and H_2S concentration depletion in the gas phase along fiber length for methanol absorbent at $V_L = 1$ m/s, $V_G = 3$ m/s and T = 298 K.

Note that because the model results are based on "nonwetted mode" in which the gas mixture filled the membrane pores and assuming complete non-wetting conditions is valid at low pressure operations, it is expected that model overestimates CO_2 removal efficiency at high pressure. This can be attributed to mass transfer resistance caused by partial wetting of pores in hollow fiber membrane [5].

The membrane length needed to achieve the desired removal efficiency is a significant parameter. By increasing the membrane length, the membrane area for mass transfer increased and thus, higher removal efficiency is achieved. As shown in Figure 13, we examined this effect for two membrane length, i.e. 27 cm and 54 cm. As a result of doubling the length, CO_2 removal efficiency increased by 60% and H_2S removal efficiency by 40% for $V_L = 1$ m/s, $V_G = 3$ m/s, $C_{0,CO_2=}$. $C_{0,H_2S} = 4$ mol/m^3 and T = 298 K.

CONCLUSIONS

The physical absorption of CO_2 and H_2S from $CO_2/H_2S/CH_4$ mixture (when the partial pressure of components is 10% of total pressure) was simulated. The effect of liquid velocity, gas velocity, temperature and pressure on removal efficiency was explored and the concentration distributions inside the shell, through the membrane, and within the tube side were studied. The results indicate that methanol has the potential as a low-cost, green physical solvent for CO_2 capture in HFMGA. Relative absorption rate of CO_2 using methanol absorbent is in the range 1.5 to 2.7 and relative absorption rate of H_2S using methanol absorbent is in the range 2.9 to 3.75 in comparison with the case of water absorbent for given operating conditions ($V_G = 2$ m/s, $C_{0,CO_2=} = C_{0,H_2S} = 4$ mol/m^3, T = 298 K). However, in simultaneously absorption of CO_2 and H_2S using methanol in MGA selectivity remains relatively constant with increasing the methanol absorbent flow rate. With increasing the temperature, the removal efficiencies slightly decreased. At high pressures, methanol as a physical solvent is more efficient and it might be an alternative to chemical solvents. Moreover, CO_2 removal efficiency about 60% and H_2S removal efficiency about 40% increased as a result of doubling the membrane length.

COMMEMORATION

The authors are particularly grateful to Prof. Mohammad Khoshnoodi from University of Sistan and Baluchestan, the project manager who passed away unexpectedly, for his suggestions and knowledge shared.

REFERENCES

1.   R. Faiz and M. Al-Marzouqi, "Mathematical Modeling for the Simultaneous Absorption of CO_2 and H_2S Using MEA in Hollow Fiber Membrane Contactors," Journal of Membrane Science, Vol. 342, No. 1-2, 2009, pp. 269-278. doi:10.1016/j.memsci.2009.06.050

2.   R. N. Maddox, "Gas Conditioning and Processing," Campbell Petroleum Series, 3rd Edition, Vol. 4, 1982.

3.   L. Sumin, et al., "The Enhancement of CO_2 Chemical Absorption by K_2CO_3 Aqueous Solution in the Presence of Activated Carbon Particles," Chinese Journal of Chemical Engineering, Vol. 15, No. 6, 2007, pp. 842-846. doi:10.1016/S1004-9541(08)60012-9

4.   A. Mandowara and P. K. Bhattacharya, "Membrane Contactor as Degasser Operated under Vacuum for Ammonia Removal from Water: A Numerical Simulation of Mass Transfer under Laminar Flow Conditions," Computers and Chemical Engineering, Vol. 33, No. 6, 2009. pp. 1123-1131. doi:10.1016/j.compchemeng.2008.12.005

5.   A. Gabelman and S. T. Hwang, "Hollow Fiber Membrane Contactors," Journal of Membrane Science, Vol. 159, No. 1-2, 1999, pp. 61-106. doi:10.1016/S0376-7388(99)00040-X

6.   V. Y. Dindore, D. W. F. Brilman and G. F. Versteeg, "Modelling of Cross-Flow Membrane Contactors: Mass Transfer with Chemical Reactions," Journal of Membrane Science, Vol. 225, No. 1-2, 2005, pp. 275-289. doi:10.1016/j.memsci.2005.01.042

7.   H. Jeon, et al., "Absorption of Sulfur Dioxide by Porous Hydrophobic Membrane Contactor," Desalination, Vol. 234, No. 1-3, 2008, pp. 252-260.doi:10.1016/j.desal.2007.09.092

8.   K. A. Hoff, et al., "Modeling and Experimental Study of Carbon Dioxide Absorption in Aqueous Alkanolamine Solutions Using

a Membrane Contactor," Industrial & Engineering Chemistry Research, Vol. 43, No. 16, 2004, pp. 4908-4921. doi:10.1021/ie034325a

9.   A. Kohl and R. Nielsen, "Gas Purification," 5th Edition, Gulf Publishing Company, Houston, 1997.

10.   G. Hochgesand, "Rectisol and Purisol," European and Japanese Chemical Industrials Symposium, 1970, Vol. 62, No. 7, pp. 37-43.

11.   J. A. Delgado, et al., "Simulation of CO_2 Absorption into Aqueous DEA Using a Hollow Fiber Membrane Contactor: Evaluation of Contactor Performance," Chemical Engineering Journal, Vol. 62, 2009, pp. 396-405. doi:10.1016/j.cej.2009.04.064

12.   R. Wang, D. F. Li and D. T. Liang, "Modeling of CO_2 Capture by Three Typical Amine Solutions in Hollow Fiber Membrane Contactors," Chemical Engineering and Processing, Vol. 43, No. 7, 2004, pp. 849-856. doi:10.1016/S0255-2701(03)00105-3

13.   W. Rongwong, R. Jiraratananon and S. Atchariyawut, "Experimental Study on Membrane Wetting In Gas-Liquid Membrane Contacting Process for CO_2 Absorption by Single and Mixed Absorbents," Separation and Purification Technology, Vol. 69, 2009, pp. 118-125.doi:10.1016/j.seppur.2009.07.009

14.   D. Wang, W. K. Teo and K. Li, "Removal of H_2S to Ultra-Low Concentrations Using an Asymmetric Hollow Fibre Membrane Module," Separation and Purification Technology, Vol. 27, No. 1, 2002, pp. 33-40. doi:10.1016/S1383-5866(01)00186-1

15.   P. Keshavarz, J. Fathhikalajahi and S. Ayatollahi, "Mathematical Modeling of the Simultaneous Absorption of Carbon Dioxide and Hydrogen Sulfide in a Hollow Fiber Membrane Contactor," Separation and Purification Technology, Vol. 63, No. 1, 2008, pp. 145-155. doi:10.1016/j.seppur.2008.04.008

16.   S. Wang, K. Hawboldt and M. A. Abdi, "Novel DualMembrane Gas-Liquid Contactors: Modelling and Concept Analysis," Industrial & Engineering Chemistry Research, Vol. 45, No. 23, 2006, pp. 7882-7891. doi:10.1021/ie051368d

17.   A. Mansourizadeh, A. F. Ismail andT. Matsuura, "Effect of Operating Conditions on the Physical and Chemical CO_2 Absorption through the PVDF Hollow Fiber Membrane Contactor," Journal

of Membrane Science, Vol. 353, No. 1-2, 2010, pp. 192-200. doi:10.1016/j.memsci.2010.02.054

18. R. Faiz and M. Al-Marzouqi, "CO$_2$ Removal from Natural Gas at High Pressure Using Membrane Contactors: Model Validation and Membrane Parametric Studies," Journal of Membrane Science, Vol. 365, No. 1-2, 2010, pp. 232-241.doi:10.1016/j.memsci.2010.09.004

19. J. Happel, "Viscous Flowrelative to Arrays of Cylinders," AIChE Journal, Vol. 5, No. 2, 1959, pp. 174-177. doi:10.1002/aic.690050211

20. M. Al-Marzouqi, et al., "Modeling of CO$_2$ Absorption in Membrane Contactors," Separation and Purification Technology, Vol. 59, No. 1, 2008, pp. 286-293.

21. M. Mavroudi, S. P. Kaldis and G. P. Sakellaropoulos, "Reduction of CO$_2$ Emissions by a Membrane Contacting Process," Fuel, Vol. 82, No. 15-17, 2003, pp. 2153-2159.doi:10.1016/S0016-2361(03)00154-6

22. P. Keshavarz, J. Fathhikalajahi and S. Ayatollahi, "Analysis of CO$_2$ Separation and Simulation of a Partially Wet-Ted Hollow Fiber Membrane Contactor," Journal of Hazardous Materials, Vol. 152, No. 3, 2008, pp. 1237-1247. doi:10.1016/j.jhazmat.2007.07.115

23. G. F. Versteeg and W. P. M. Van Swaaij, "On the Kinetics between CO$_2$ and Alkanolamines both in Aqueous and Non-Aqueous Solutions I. Primary Andsecondary Amines," Chemical Engineering Science, Vol. 43, No. 3, 1988, pp. 573-585. doi:10.1016/0009-2509(88)87017-9

24. J. J. Carroll and A. E. Mather, "The Solubility of HydroGen Sulphide in Water from 0°C to 90°C and Pressure to 1 MPa," Geochimica et Cosmochimica Acta, Vol. 53, No. 6, 1989, pp. 1163-1170. doi:10.1016/0016-7037(89)90053-7

25. K. Lunsford and G. Mcintyre, "Decreasing Contactor Temperature Could Increase Performance," GPA Annual Convention, Bryan Research and Engineering, Inc., Texas, 1999, pp. 121-127.

26. E. L. Cussler, "Diffusion Mass Transfer in Fluid Systems," Cambridge University, Cambridge, 1984.

27. M. V. Diaz and A. Coca J., "Correlation for the Estimation of Gas-Liquid Diffusivity," Chemical Engineering Communications, Vol. 52, No. 4-6, 1987, pp. 271-281.doi:10.1080/00986448708911872

28.   R. H. Perry, "Perry's Chemical Engineers' Handbook," 7th Edition, McGraw-Hill, New York, 1997.

29.   B. E. Poling, J. M. Prausnitz and J. P. O'Connell, "The Properties of Gases and Liquids," 5th Edition, McGrawHill, New York, 2004.

30.   R. Faiz and M. Al-Marzouqi, "H_2S Absorption via CarBonate Solution in Membrane Contactors: Effect of Species," Journal of Membrane Science, Vol. 350, No. 1-2, 2010, pp. 200-210. doi:10.1016/j.memsci.2009.12.028

31.   V. Y. Dindore, D. W. F. Brilman and G. F. Versteeg, "Hollow Fiber Membrane Contactor as a Gas-Liquid Model Contactor," Chemical Engineering Science, Vol. 60, No. 2, 2005, pp. 467-479. doi:10.1016/j.ces.2004.07.129

Chapter 5

The Wear Analysis Model of Drill Bit Cutting Element with Torsion Vibration

Jialin Tian[1, 2], Chuanhong Fu[1], Lin Yang[1], Zhi Yang[1], Xiaolin Pang[1], You Li[1], Yonghao Zhu[1], and Gang Liu[1]

[1]School of Mechanical Engineering, Southwest Petroleum University, Chengdu 610500, China
[2]School of Mechanical Engineering, Southwest Jiaotong University, Chengdu 610031, China

ABSTRACT

Due to some deficiencies in the existing researches of cutting element wear of polycrystalline diamond compact (PDC) drill bit and to reduce the wear rate of cutting element of PDC bit, accurately illustrate the effect of tooth-distributing angle on cutting element, and improve the service life of PDC drill bit, we use the geometry knowledge of PDC drill bit and establish the torsion vibration model of PDC drill bit, solve and analyze the model, and at last analyze the effect of tooth-distributing angle on the rule of cutting element wear by using the wear theory.

INTRODUCTION

The results of polycrystalline diamond compact (PDC) drill bit rock breaking are affected by many factors, including the weight on bit (WOB), rotation speed, torque, which are related to load, and the cutters' density and tooth-distributing angle, which related to structure. According to the coupling relationship between cutters and rock, the failure modes of cutters (as cutting element) are as follows: the rock-breaking volume of blade's central cutting element is small, the main failure mode is wear, and the wear amount is small, while the linear velocity of blade's peripheral cutting element is big and the effective volume of rock breaking is big, and the normal failure mode is wear and the abnormal failure mode is crack. Additionally, the failure mode of drill bit's gauge tooth is also wear.

Aiming at the existing researches for wear failure of PDC drill bit's cutting element by experts and scholars at home and abroad, the literature [1] thought the transient of PDC drill bit's processing technology (tool cutting, fast cooling, and pressure removal) was a major reason for the formation of microcrack and led to premature failure of the drill bit. The literature [2] described the structure and parameter designs of PDC drill bit in depth and how to improve the service life and reduce the wear rate of PDC drill bit. According to some materials used by PDC drill bit, the literature [3] carried out a detailed study and the author believed that the main reasons for the failure of cutting element were wear resistance and poor impact toughness of WC-Co carbide which could not withstand harsh working environment at bottom hole. The literature [4] thought, in the process of drilling, the vibration of drill string, especially transverse vibration, was the root of premature failure of PDC drill bit. The literature [5] thought the failure of PDC drill bit was mainly due to repeated impact load which was the main reason for crack initiation and growth and was ultimately leading to failure.

Due to the difference between the above researches and the real failure, the paper has studied the blade's peripheral cutting element with the biggest wear amount. Based on the dynamics and tribology, we have analyzed the effect of the drill bit's tooth-distributing angle on the law of cutting element wear. Given the WOB, speed, and torque as load parameters and the bit blade number and cutters' density as structure parameters, we have established the mathematical model of the wear rule affected by tooth-distributing angle, which is based on the dynamic wear theory. The load in the analysis model is the dynamic load of bit drilling process, which is more consistent with the cutting element's real load in the rock-breaking process, and more accurately reflects the effect of tooth-distributing angle on the dynamic wear rule. The results of the model can obtain the quantitative indicators which have a guiding significance to the optimization on tooth distribution of PDC drill bit.

CALCULATION MODEL

Rock cutting by PDC drill bit's cutting teeth at bottom hole is actually a problem of cutting element's oblique cutting. As for PDC drill bit, the geometry model is the foundation of the study of the cutting mechanics and dynamics of PDC drill bit, especially in which the changes in the parameters of tooth-distributing angle have a great effect on the force and wear resistance of PDC drill bit. The geometry model is in Figure 1.

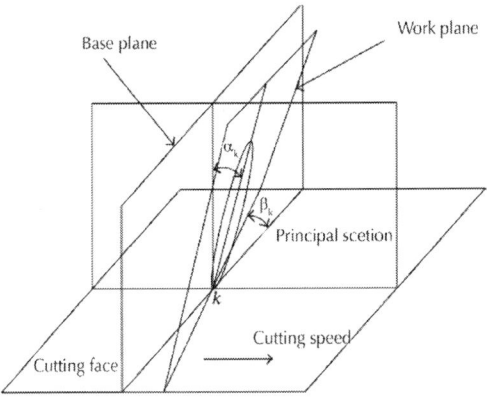

Figure 1: Geometry model of PDC drill bit's cutting teeth.

According to the definition of PDC drill bit's geometryangle, assume the coordinate of any point of cutting element(cutter) as $w(r, \psi, h_c)$, where r is the radius of cutter, ψ is the location angle of cutter, h_c is the axial distance of cutter's anchor point, ξ is the normal angle of cutter, α is the cutting edge angle of cutter, β is the side rake angle of cutter. According to the analysis and accurate calculation of PDC cutter working angle in the article by Li et al. [6], the normal direction vector of the cutter working surface can be obtained:

$$U = \{u, v, p\},$$

(1)

where

$$u = \cos\alpha\cos\beta\cos\psi + \sin\alpha\sin\psi$$

$$v = \cos\alpha\cos\beta$$

$$p = \sin\alpha\cos\psi - \cos\alpha\sin\beta\sin\psi.$$

(2)

According to the relationship of the geometric model and coordinate conversion, we have established the computational formula of PDC cutting element's cutting edge angle, front rake angle, side rake angle, and back rake angle. The geometry formula [7] of the cutting edge angle and back rake angle is obtained as

$$\beta_k = \arctan\frac{u\left[\zeta_1\left(r + x_1\right) + \zeta_2 y_1\right] + w\left[\zeta_3\sqrt{\left(r + x_1\right)^2 + y_1^2}\right]}{v\sqrt{2\left[\zeta_1\left(r + x_1\right) + \zeta_2 y_1\right]^2}},$$

(3)

where β_k is the cutting edge angle, rad, and the computational formulas of $x_1, y_1, \zeta_1, \zeta_2, \zeta_3$ are as

$$x_1 = r\sin\psi\cos\beta\cos\xi$$

$$+ r\cos\psi\left(\cos\alpha\sin\xi - \sin\alpha\sin\beta\cos\xi\right)$$

$$y_1 = -r\sin\psi\sin\beta - r\cos\psi\sin\alpha\cos\beta$$

$$\zeta_1 = r\left(\cos\psi\cos\beta\cos\xi + \sin\psi\sin\alpha\sin\beta\cos\psi\right)$$

$$\zeta_2 = r\left(\sin\psi\sin\alpha\cos\beta - \cos\psi\sin\beta\right)$$

(4)

As shown in Figure 2, assume any point in PDC drill bit's cutting element as (x, y), when 0<y<r; the computational formula of the cutting point is as

$$x = \left(r^2 - \left(r - \frac{a_c}{\cos \alpha} \right)^2 \right)^{1/2}$$

$$y = \frac{a_c}{\cos \alpha} \quad \text{for } 0 < y < r,$$

(5)

where a_c is the cutting depth of compact, mm.

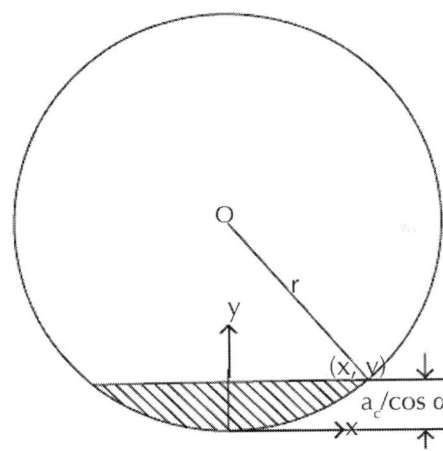

Figure 2: Diagram of PDC drill bit's cutting element.

According to the computational formulas of geometry, the formulas of the effective cutting edge length l_{ccl}, cutting arc length L, and cutting area $A_{cutting}$ can be obtained as

$$l_{ccl} = 2x$$

$$L = 2r \arccos \left(1 - \frac{y}{r} \right) \cos \beta$$

$$A_{cutting} = \left(r^2 \arccos \left(1 - \frac{y}{r} \right) \right.$$

$$\left. - r(r - y) \sin \left(\arccos \left(1 - \frac{y}{r} \right) \right) \right) \cos \beta.$$

(6)

Substituting formulas (5) into formula (6), we can obtain the calculations of the cutting edge cutting, cutting arc length, and cutting area.

In the drilling process of PDC drill bit, the motion of cutting element at bottom hole is extremely complex. Assuming the bit axis coincides with the well center line, the cutting teeth have two main directional motions at bottom hole; one is the rotation motion of cutting teeth round the bit axis, and the other is the downward motion of cutting teeth together with the bit. To facilitate the analysis of torsional vibration of PDC drill bit, we have made some assumptions: the drill string and drill collar are both homogeneous elastomers; the rotation speed of the top swivel plate and the WOB always remain the same. The simplified model of PDC drill bit with dynamics of torsional vibration is in Figure 3.

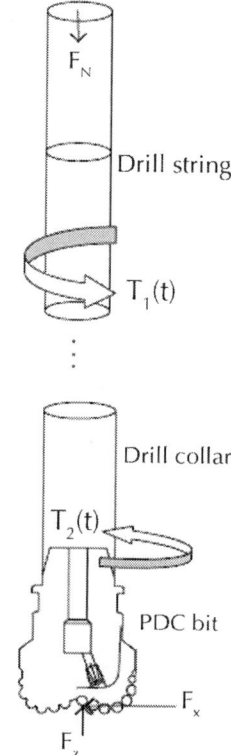

Figure 3: Dynamic model of PDC drill bit.

Through the dispersed method, we transform the established dynamic system of PDC drill bit with torsional vibration into a discrete system of a limited degree freedom; the differential vibration equation of PDC drill bit can be written as the following matrix form:

$$[M] \left\{ \ddot{\theta} \right\} + [C] \left\{ \dot{\theta} \right\} + [K] \left\{ \theta \right\} = [T(t)],$$

(7)

where $[M]$ is the whole moment of inertia matrix, $[C]$ is the whole damping matrix, $[K]$ is the whole stiffness matrix, $[T(t)]$ is the whole dynamic torque matrix, $\{\theta\}$ is the whole angle matrix, $\{\dot{\theta}\}$ is the whole angle velocity vector, and $\{\ddot{\theta}\}$ is the whole rotational acceleration vector.

In the drilling process, the torsional vibration of drill string and drill collar makes PDC drill bit produce very high dynamic torque at bottom hole. The dynamic torque imposed on PDC drill bit by drill string is $T_1'(t)$; at this point the dynamic torque and drill string's positive torque $T_1(t)$ on swivel plate are the same, while rock will produce resistance torque $T_2(t)$ to PDC drill bit; at this point the dynamic torque $T(t)$ on the drill can be expressed as

$$T(t) = T_1'(t) - T_2(t).$$

(8)

According to formulas (7) and (8), we can work out the rotation angle θ (rad) and rotation angular velocity $\dot{\theta}$ (rad/s) of PDC drill bit that rotates around the axis at bottom hole. Due to the relationship between the linear velocity and angular velocity, we can calculate the linear velocity V_{rl} (mm/s) of PDC drill bit at bottom hole. The specific formula is as

$$V_{rl} = r\dot{\theta}.$$

(9)

The motion of PDC drill bit at bottom hole is partly from the downward motion of the bit; namely, the axial linear velocity V_{al} (mm/s) of the drill bit at bottom hole is as

$$V_{al} = \frac{1000 R_j}{3600},$$

(10)

where R_j is the rate of penetration (ROP), (m/h).

At this point, the absolute linear velocity V of PDC drill bit at bottom hole is

$$V = \left(V_{rl}^2 + V_{al}^2 \right)^{1/2}.$$

(11)

According to the relationship between rotation speed and linear velocity, the rotation speed of PDC drill bit at bottom hole can be calculated as

$$N = \frac{V}{2\pi r},$$

(12)

where N is the rotation speed of PDC drill bit, (rad/min).

In oil drilling, the failure of PDC cutting element due to normal wear accounts for most of its failure. According to the principle of nonlinear dynamical system, the wear process can be divided into three stages, namely, the stages of self-organization, chaos, and instability, corresponding to the three stages of running-in wear, normal wear, and sharp wear in the expression of tribology. If the effect of dynamic on the wear of PDC compact had not considered, it would not reflect the true wear situation of PDC bit. As a result, the wear equation based on torsional vibration is very important for the study of cutting element wear.

In the drilling process of PDC drill bit, we assume the time of cutting rock as t, substitute the drill speed obtained by formula (12) into the wear volume formula, and obtain that [8]

$$V = \tau \pi l \left(l + 2\delta \right) \mu p N t,$$

(13)

where τ is the cutting element wear strength, N/m^2, p is the pressure distribution density, N/m, l is the effective cutting edge length of cutting element, mm, μ is the friction coefficient between cutting element and rock, and δ is the distance from wear parts to the center of drill bit, mm.

The pressure distribution density p in formula (13) can also be expressed by the following formula:

$$p = \frac{\exp\left(a_c/D\cos^2\phi\right) F_d}{L},$$

(14)

where F_d is the PDC axial pressure, N, L is the arc length produced when the cutting element cuts rock, mm, D is the diameter of PDC drill bit, mm, and ϕ is the caster angle, (°).

Substituting formula (14) into formula (13) and dividing by time t, we obtain the cutting element's volume wear rate by the tooth-distributing angle of PDC drill bit based on torsion vibration:

$$v_t = \pi l\,(l + 2\delta)\,C_p\mu\tau N \exp\left(\frac{a_c}{D\cos^2\phi}\right)\frac{F_d}{L},$$

(15)

where v_t is the cutting element's volume wear rate, mm³/h, and C_p is the hydraulic coefficient.

In the design of PDC drill bit, with the increase of the diameter, the wear linear velocity will increase, and the rock-breaking volume of the cutter is also on the rise. Assuming that the cutting depth is 1 mm and substituting formula (12) into formula (15), we obtain the relationship between the linear velocity, diameter, and volume wear rate:

$$v_t = \pi l\,(l + 2\delta)\,C_p\mu\tau\frac{V}{D\pi} \exp\left(\frac{1}{D\cos^2\phi}\right)\frac{F_d}{L}.$$

(16)

Integrating formula (16), we obtain the computational formula of the back rake angle:

$$a = \arccos\left(\frac{1}{r\left(1 - \cos\left(\pi l\,(l + 2\delta)\,C_p\mu\tau N \exp\left(1/D\cos^2\phi\right) F_d/2r\cos\beta v_t\right)\right)}\right).$$

(17)

THE ANALYSIS OF EXAMPLES

Using the geometry equation of drill bit's tooth-distributing angle and according to the given example parameters, we have calculated and analyzed the relationship between the cutting edge angle and back rake angle, as well as the effect of the back rake angle on the cutting area and cutting arc. Then we have analyzed the dynamics of PDC drill bit in the process of drilling and put the solved bit speed into the solving theory of cutting element's wear rate to analyze wear. The partial example parameters are in Table 1 and the results are in Table 2, Figures 4 and 5.

Table 1: Specific parameters

Parameter name	Results
Side rake angle β (°)	9
Cutting depth a_c (mm)	1
Cutting tooth position angle ψ (°)	11
Cutting tooth normal angle ξ (°)	9.5
PDC drill bit type	8.5"
Blade number (a)	6
Composite pad (main cutter) radius r (mm)	7

Table 2: Calculations of the back rake angle, cutting edge angle, and effective length of cutting edge

Back rake angle α (°)	10	12	14	16	18	20	22	24	26	28
Cutting edge angle β_k (rad)	0.0134	0.0135	0.0135	0.0136	0.0137	0.0138	0.0138	0.0139	0.0140	0.0141
Effective length of the cutting edge l (mm)	7.8014	7.8261	7.8555	7.8897	7.9290	7.9733	8.0289	8.0781	8.1392	8.2063
Cutting arc L (mm)	7.9012	7.8739	7.8416	7.8042	7.7619	7.7144	7.6557	7.6044	7.5416	7.4736
Cutting area $A_{cutting}$ (mm²)	4.7732	4.6057	4.4074	4.1781	3.9178	3.6261	3.2648	2.9490	2.5621	2.1430

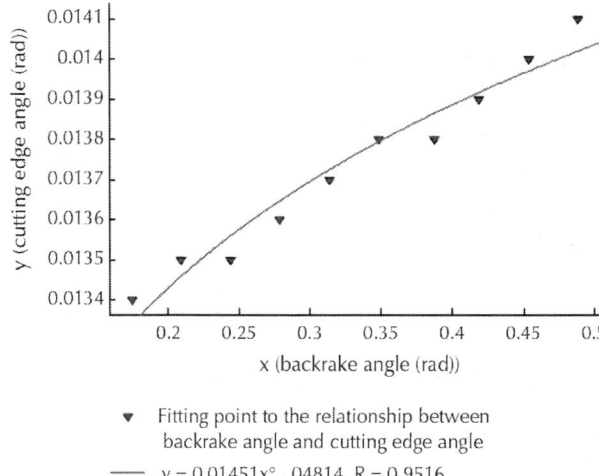

Figure 4: Relationship between the cutting edge angle and the back rake angle.

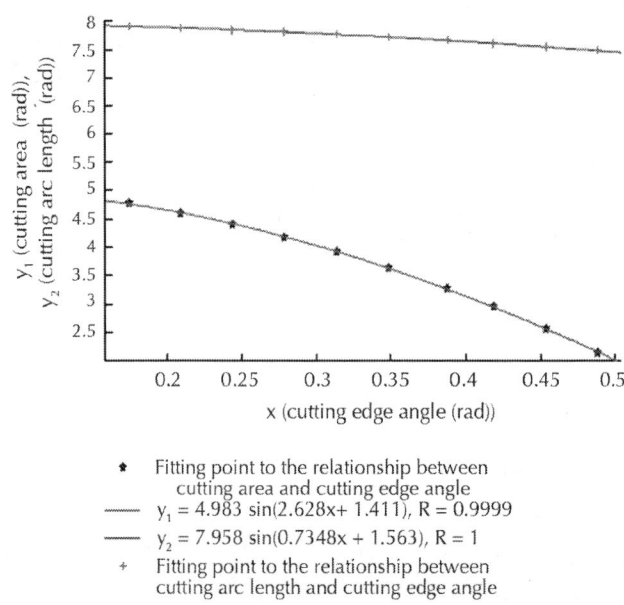

Figure 5: Relationship between the cutting edge angle and the cutting area, cutting the arc length.

Through the analysis of Figures 4 and 5, there is a close relationship between the back rake angle, cutting edge angle, effective cutting edge length, cutting area, and cutting arc length. From the values of the back rake angle and cutting edge angle, with the increase of back rake angle, the cutting edge angle and effective length of cutting edge both increase, but we found that the increase of cutting edge angle is not obvious; namely, the increasing rate of cutting edge angle is about 300 times that of cutting edge angle, and known by the way of curve fitting, they have a relationship of an exponential function. With the increase of the cutting edge angle, the cutting area and cutting arc length decrease obviously; namely, the tiny change of back rake angle has a great effect on them. What is more, the decreasing speed of the cutting area is far greater than that of the cutting edge angle; namely, the change of the back rake angle has a greater effect on the cutting area. By the curve fitting, we find that the above figure curves all accord with the distribution characteristic of sine function, but seen from the fitting accuracy, the fitting between the cutting edge angle and cutting length is more accurate. Due to the increase of the back rake angle, it makes the effective cutting edge length increase and the cutting area and cutting arc length decrease, which significantly reduces rock-breaking efficiency and ROP. As a result, it will have no high efficiency of rock breaking.

According to the computational formula of dynamics and the specific parameters in Table 3, the rotation angle and vibration angular velocity of PDC drill bit at bottom hole can be obtained in the process of drilling, as well as the speed of PDC drill bit under the common action of the rotary motion and downward movement, as shown in Figures 6, 7, and 8.

Table 3: Specific parameters

Parameter name	Results
Active torque $T_1(t)$ (N·m)	20000
Resistance torque $T_2(t)$ (N·m)	9000
Drill pipe unit mass (kg/m)	9.32
Drill collar unit mass (kg/m)	36.99
Damping ratio	0.1

Drill pipe inner diameter (mm)	46.1
Drill pipe outer diameter (mm)	60.3
Drill collar inner diameter (mm)	35
Drill collar outer diameter (mm)	85
Mechanical drilling rate R_j (m/h)	2.4
Well depth (m)	1000

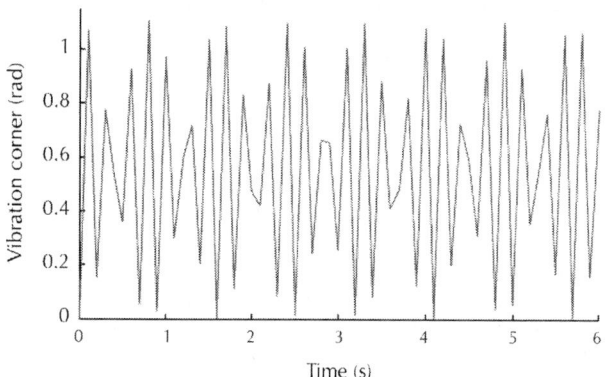

Figure 6: Vibration corner of PDC drill bit.

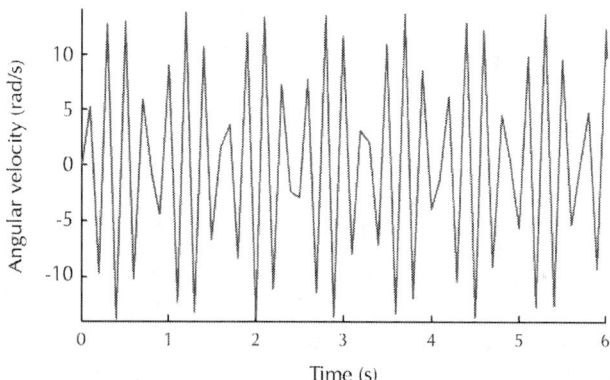

Figure 7: Vibration angular velocity of PDC drill bit.

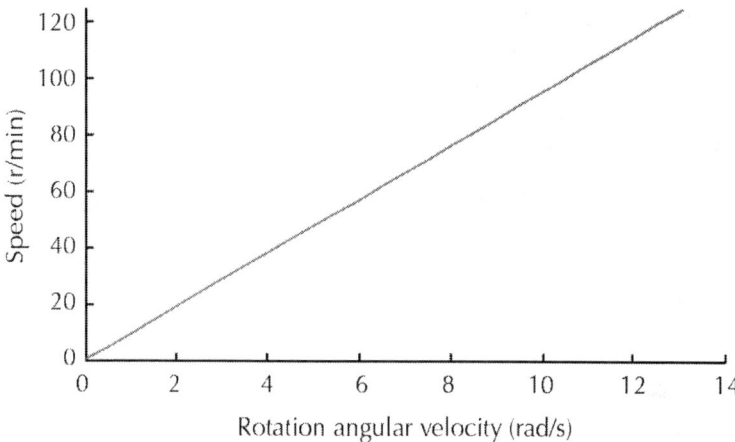

Figure 8: Relationship between PDC drill bit's rotation angular velocity and speed.

From Figure 8 the following is shown: with the increase of angular velocity, the speed of PDC drill bit is also on the rise, basically in a linear distribution. And in the process of drilling, the strong vibration of PDC drill bit at bottom hole will have a great effect on the PDC drill bit performance, service life, tooth-distributing angle of PDC drill bit, and cutting element wear.

According to formula (15), we can cognize the effect of PDC drill from the center to the periphery of blade on the wear rule of cutting element. Assume the cutting arc length as 7.9012 mm; the result is shown in Figure 9. By determining the volume wear rate of each part of cutting element, we can obtain the back rake angle and cutting edge angle, which provide the reference for the optimization of tooth distribution. At the same time, we also need to consider the tooth-distributing blade numbers, diameter, and other parameters together to achieve the purpose of optimizing the rock-breaking performance. Taking the cutting element at the gauge diameter, for example, we assume that the WOB is 80 KN; the axial pressure that is distributed in each cutting element and the other parameters used in computational analysis are shown in Table 4.

Table 4: Specific parameters

Parameter name	Results
PDC axial pressure F_d (N)	390
Compact wear strength τ (N/mm²)	70×10^6
Hydraulic coefficient C_p	1.35×10^{-9}
Friction coefficient between cutting element and rock μ	0.38
Caster angle ϕ (°)	12

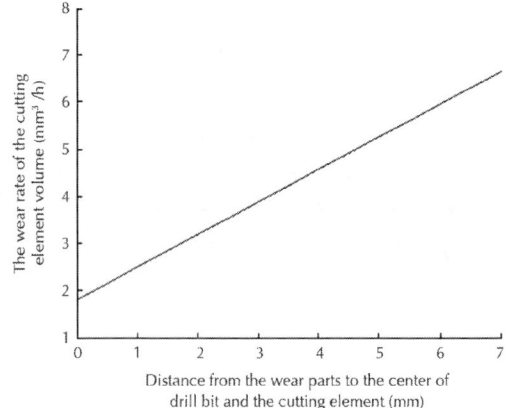

Figure 9: Distance from the wear parts to the center of drill bit and the cutting element.

In Figure 9, the cutting element's volume wear rate v_t and the distance δ from the wear parts to the center of drill bit have a direct relationship; with the increase of δ, v_t is also on the rise, which is in a linear distribution. This phenomenon shows that the rock-breaking volume and wear amount of bit's central cutting element are small, while those of blade's peripheral cutting element are large. As a result, the study of largest wear amount of blade's peripheral cutting element is reasonable in this paper.

Substituting the data in Table 4 into formula (16), we get the calculations of cutting element's wear rule affected by the back rake angle, shown in Figure 10.

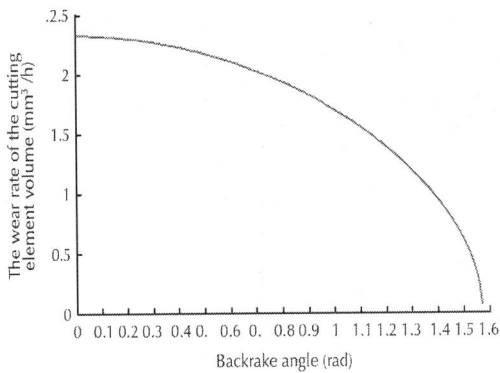

Figure 10: Relationship between the back rake angle and wear rate of cutting element volume.

According to the relationship between the back rake angle and front rake angle, the above listed data and the formula of the volume wear rate of PDC cutting element based on torsion vibration, we can obtain the result of cutting element's wear rule affected by the cutting edge angle, shown in Figure 11.

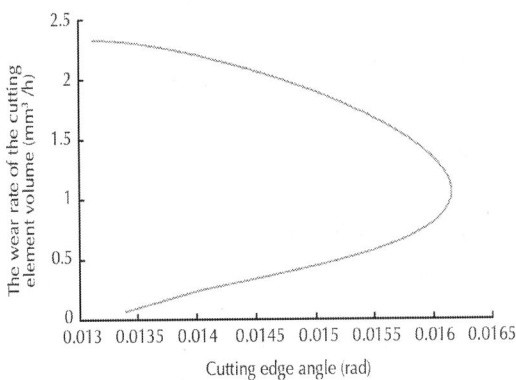

Figure 11: Relationship between the cutting edge angle and wear rate of cutting element volume.

According to the results of Figures 10 and 11, in the analysis of cutting element's wear rule affected by tooth-distributing angle, the value of tooth-distributing angle ranges from 0° to 90°, because the selection of the angle is not greater than 90° in the actual processing application. With the increase of the back rake angle of PDC drill bit, the cutting element's volume wear rate decreases significantly. The cutting edge angle has a very complex effect on the volume wear rate of PDC cutting element. Seen from Figure 11, with the increase of cutting edge angle, the volume wear rate of cutting element decreases, but the decreasing trend is nonlinear. At the same time, combined with Figure 3, with the increase of back rake angle, the cutting edge angle is also on the increase, when the inclination increases and the cutting element's volume wear rate decreases, and the increase of the cutting edge angle is also bound to decrease, verifying the correctness of Figure 11. As a result, we should choose a large tooth-distributing angle to reduce the volume wear rate of compact, which improves service life of drill bit, but the unduly large back rake angle will change the cutting direction and reduce cutting efficiency and service life of bit, so how to choose a suitable tooth-distributing angle will have a great effect on the performance of PDC drill bit and its service life.

FIELD DATA COMPARATIVE ANALYSIS

Calculation Results and Experimental Verification

According to the corresponding numerical parameters, results, and optimization of tooth-distributing parameters by each performance index from drill design, we have processed four experimental bits with different tooth-distributing angles as shown in Table 5. In order to verify the correctness of calculations better, we have done experiments in the third section of drill 1 well and second section of drill 2 well in an oil field where the wellbore is 311.2 mm, the drilling rig is ZJ70J, the mud pump is F-1600, the cylinder liner diameter is 150 mm, the drilling fluid density is 1.02, the viscosity is 60 s, the sediment concentration is 0.3%, the new degrees of the four experimental bits are 100% and 30%, respectively, when the four drills go into and out of the wells; the other specific experimental parameters are in Table 6. In the field test, we record ROP at all times and use the laboratory equipment including the vibration meter, sensor, and speedometer to test PDC drill bit and then import the test data into the special software, deal with the data, and analyze the speed of PDC drill bit and cutting element's volume wear rate. The experiment results are in Figures 12 and 13.

Table 5: Experimental data of drill bit

Drill number	Back rake angle (rad)	Cutting edge angle (rad)
1	0.1744	0.0134
2	0.314	0.0137
3	0.4186	0.0139
4	0.4884	0.0141

Table 6: Experimental parameters in the field

Drill number	Well number	Well section (m)	Experimental time	Footage (m)	Horizon	Lithology
1	1	2791–3041	2012.5	250	Pearl Chong	Sandstone
2	1	3041–3291	2012.7	250	Pearl Chong	Shale
3	2	2103–2353	2012.3	250	Pearl Chong	Sandstone
4	2	2323–2603	2012.5	250	Pearl Chong	Shale
Drill number	Pump pressure (MPa)	WOB (KN)	Net drilling time (h)	Discharge (L/m)	Rotary speed (rmp)	Average ROP (m/h)
1	22–31	74–82	96.1	45	124	2.6
2	22–26	74–79	108.6	45	123	2.3
3	21–32	72–84	119	45	119	2.1
4	23–26	76–79	104.1	45	122	2.4

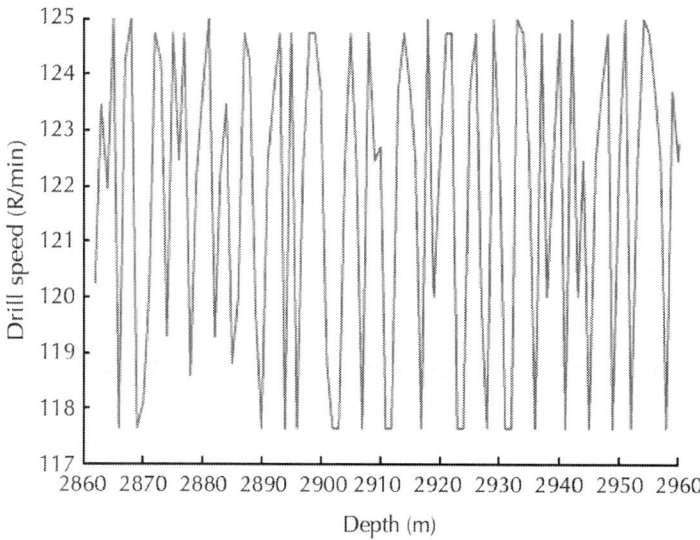

Figure 12: Speed of PDC drill bit.

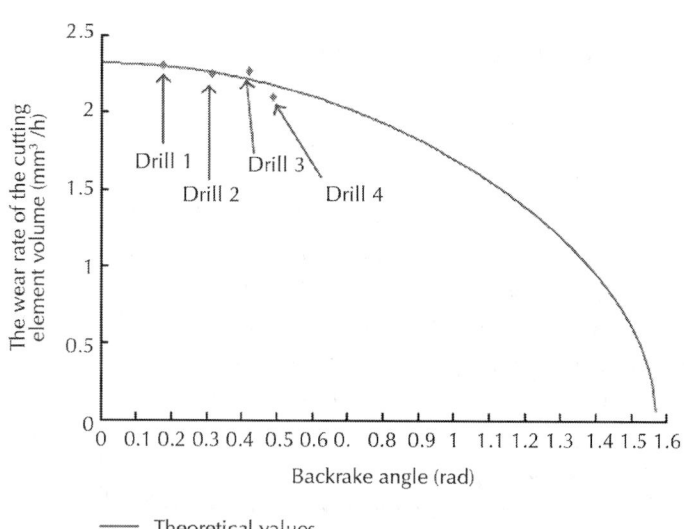

(a)

(a)

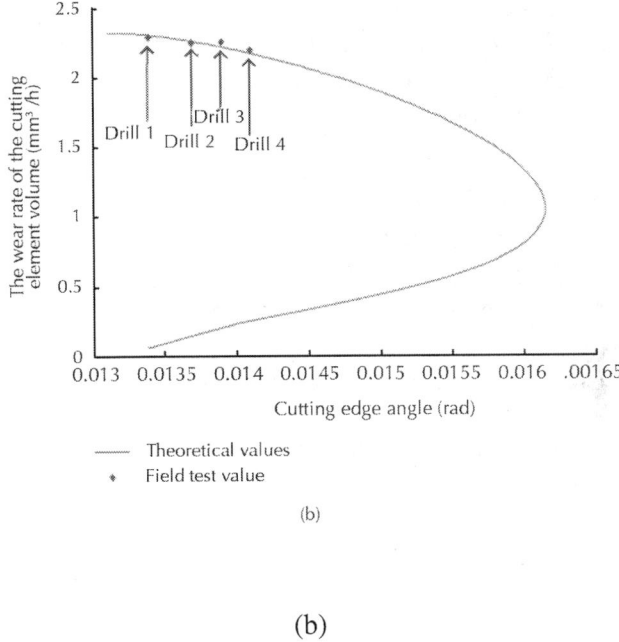

(b)

Figure 13: Comparison chart between the calculated and testing values.

Comparative Analysis

We can know, through the experiment, the lithology is mainly composed of sandstone and shale; the main compositions of sandstone are quartz with rich black mineral, grain structure, mediocre separation, subrounded particle, argillaceous cement, densification, and hard characteristic; shale contains part of silty clay with developing lamellation, densification, and brittle and medium-hard characteristic.

According to the degree of drill wear, we find that drill 1 drill wear is the most serious, the gauge teeth have wear, the drill teeth wear is very serious, and the outer row of teeth has many broken teeth. Drill 4 wear is the lightest, the gauge teeth basically have no wear, the drill teeth wear a little, and the outer row of teeth has no broken teeth. Drill 2 and drill 3 are worse than drill 4; the outer row of teeth has the phenomenon of broken teeth, but compared with drill 2, drill 3 drill wear is worse and the phenomenon of broken teeth of the outer row of teeth is a little more.

We get that the speed and cutting element's volume wear rate of PDC drill bit, the same as the drill bit in the calculation parameters, basically agree with the experimental results which verify the rationality and accuracy of the theoretical calculation.

The test speeds of the PDC drill bit in the process of the experiment are not stable, but a certain degree of vibration is normal, because the physical properties, chemical properties, and soft and hard degree of rock are different, which will certainly produce different degrees of resistance to PDC cutting element, so it makes PDC drill bit have a certain degree of instability in the bottom-hole movement.

We can see from Figure 13 that there are still some errors on the calculated and experimental values of the cutting element's volume wear rate of PDC drill bit affected by the back rake angle and cutting edge angle; the error range within five percent is acceptable in engineering. The error factors are various; for example, we did not take into account the effects on drill string friction by the wall in the theoretical analysis. In the process of analyzing torsion vibration, the drill bit axis is often supposed to coincide with the well centerline, but in the real drilling, it is impossible that the axis of PDC drill bit and the well centerline coincide. We take the vertical well, for example, and think that a variety of factors such as no deflection angle lead to errors in the process of drilling.

CONCLUSIONS

Compared with the existing research methods in the references which consider cutting element wear just from the point of statics or the whole view of bit, our paper has analyzed cutting element wear affected by PDC tooth-distributing angle based on torsion vibration, which makes computational accuracy improve by about 10%~20%. From each point of affecting PDC tooth-distribution, we know that the back angle and cutting edge angle have the most serious effect on PDC cutting element wear; this conclusion has certain reference significance to how to reduce wear failure of PDC cutting element.

The wear rate of cutting element has a close relationship with not only the back rake angle and cutting edge angle, but also the hard and soft degree of rock; the wear degree of drill 2 should be more than drill 3, but the result is actually on the contrary. What is more, when the drilling parameters and other experimental conditions are the same, the only change is the lithology, which exactly demonstrates that the lithology has an important effect on the degree of cutting element wear.

Due to the comparative analysis of PDC cutting element wear affected by the back rack angle and cutting edge angle, we can know that the effect of back rack angle on the wear rate of PDC cutting element is greater than the cutting edge angle, so how to choose a reasonable back rack angle becomes very important.

ACKNOWLEDGMENTS

This work is supported by Open Fund (OGE201403-05) of Key Laboratory of Oil & Gas Equipment, Ministry of Education (Southwest Petroleum University), National Natural Science Foundation of China (51074202, 11102173), and Major Cultivation Foundation of Sichuan Education Department (12ZZ003, no. 667)and Graduate Innovation Foundation of Southwest Petroleum University, SGIFSWPU.

REFERENCES

1. V. Kanyanta, S. Ozbayraktar, and K. Maweja, "Effect of manufacturing parameters on polycrystalline diamond compact cutting tool stress-state," International Journal of Refractory Metals and Hard Materials, vol. 45, pp. 147–152, 2014.

2. D. Qiong Guo and C. Hou, "Development of PDC drill bits for MWD directional drilling in underground coal mine," Procedia Earth and Planetary Science, vol. 3, pp. 440–445, 2011.

3. Y. Zhou, Z. Huang, F. Zhang et al., "Experimental study of WC-Co cemented carbide air impact rotary drill teeth based on failure analysis," Engineering Failure Analysis, vol. 36, pp. 186–198, 2014.

4. X. Zhu, Y. Liu, and H. Tong, "Analysis of reamer failure based on vibration analysis of the rock breaking in horizontal directional drilling," Engineering Failure Analysis, vol. 37, pp. 64–74, 2014.

5. V. Kanyanta, A. Dormer, N. Murphy, and A. Invankovic, "Impact fatigue fracture of polycrystalline diamond compact (PDC)

cutters and the effect of microstructure," International Journal of Refractory Metals and Hard Materials, vol. 46, pp. 145–151, 2014. View at Publisher · View at Google Scholar

6.   S.-S. Li, D.-K. Ma, and J.-K. Hou, "Precise calculation and analysis of working angles of PDC cutters,"Journal of Southwest Petroleum Institute, vol. 18, no. 4, pp. 67–72, 1996 (Chinese).

7.   Y. Li, Y. Ying-Xin, and L. Ming, "The PDC bit cutter working angle of transverse vibration drilling mode calculation," Natural Gas Industry, vol. 28, no. 3, pp. 72–74, 2008 (Chinese).

8.   Q. Guan-Zheng, Q. Da-Wei, G. Rui, Z. Xiao-Hua, and J. Zhong-Min, "Study on a swear rate model of PDC composite cutters," Complex Hydrocarbon Reservoirs, vol. 6, no. 1, pp. 62–71, 2013 (Chinese).

Chapter 6

Analysis of the Chemical Safety Facility Investment Performance in China

Kang Sun, Long Bai, and Xiaohong Li

Center for Studies of Marine Economy and Sustainable Development, Liaoning Normal University, Dalian, China

ABSTRACT

This paper adopts the accident incidence, the gross industry output value, the investment in safety facilities, and per capita wage of employment as the indexes to empirically analyze the investment performance of chemical safety facilities using time series data by VECM in China. The empirical results indicate that for China 's chemical industry, increasing investment fails to improve the short-term safety level significantly because of the offsetting behavior of workers. Over the long term, the offsetting behavior tends to disappear, and the chemical accident incidence can be decreased through increasing investment. Poor safety awareness among workers is one of the causes of accident incidences. The conclusions provide theoretical support for China to perfect chemical industry safety management.

INTRODUCTION

With the birth of chemical industry, controversy arose because of the industry's high risk to human safety and to the environment. In recent decades, hazardous chemical accidents have become a worldwide problem. Many international conventions relating to the chemical industry have been legislated, including regulations, policies, and industry management systems, such as the Rotterdam Convention (1998), the Stockholm Convention (2001), the UNEP (United Nations Environment Programme) Strategic Approach to International Chemicals Management (1998), and the EU (European Union) directive on Registration, Evaluation, Authorization and Restriction of Chemicals (REACH, 2007). Over the past 20 years, China has become a major international player in the chemical industry. Because of the rapid development of China's chemical industry, it has become the pillar industry of the national economy. Consequently, hazardous chemical accidents happen frequently, often engendering secondary disasters. Secondary disasters also pose big threats to human safety and health. The safety, health, and environmental problems caused by hazardous chemical accidents are increasing. The high frequency of chemical accidents has caused serious social problems in China . China has recently enacted more than 200 laws, administrative regulations, and departmental rules for its chemical industry. This intensification of management is unprecedented, but it still cannot stop the rise of chemical accidents year by year [1] . Thus, it is necessary to analyze the safety management effect empirically for China's chemical industry.

Most scholars used econometric analysis methods to research how the laws and regulations strengthened by safety regulatory organization influence the number of safety accidents brought by production and safety regulation effects empirically. Smith (1979) suggested that OSHA (Occupational Safety and Health Administration) policies were effective after analyzing the impact of its inspection on manufacturing injury rates over the period of 1973-1974 [2] . Gray and Scholz (1993) analyzed the industry panel data of 1979-1985 and concluded that OSHA policies reduced workplace fatalities by 22% [3] . In addition, Beck and Alford [4] (1980), Carmichael [5] (1986), and Weil [6] (1996) also deemed safety regulation was effective. Nevertheless, the result of Viscusi's [7] [8] (1979, 1992) study was that the more the regulation was promulgated, the more deaths were caused through researching the U.S. government's safety and health regulation policy. Other scholars [9] [10] found that the regulation effect was unsatisfactory due to the regulatory capture phenomenon. The extant literature about chemical safety mainly focuses on accident disposal [11] , accident statistical analysis [12] , and the domino effect [13] . The literature in relation to the cause of chemical accidents which happened frequently is relatively lacking, especially empirical analysis in China .

In this paper, an empirical test on the investment performance of chemical safety facilities will be implemented to talk about the cause of the chemical accidents happening frequently through analyzing the dynamic relationship between the investment and the performance of chemical safety facilities, using the time series data by VECM (Vector Error Correction Model).

SELECTION CRITERIA AND DATA

The general empirical analysis approach for regulation effect is to do a regression of the regulation behavior index to the regulation effect index, then test the significance and direction of the influence. Due to the particularity and complexity of the chemical industry, there are many factors that influence the chemical accident incidence. In this paper, index selection is mainly on the basis of existing literature. For instance, safety inspections and fines of factories had been selected as regulation behavior; workplace accident mortality had been selected as regulation effect by Klick and Stratmann (2003) in their research using the data provided by the OSHA [14] . Similarly, in this paper, in order to highlight the chemical accidents which occurred frequently in China , the accident incidence is selected as investment performance index. In order to avoid the contingency factors of accident incidence increases, we normalize the index by 100 million Yuan output value. As for the investment of chemical safety facilities, we choose the investment in fixed assets to denote investment in chemical safety facilities[1]. Following the Peltzman effect[2], the workers' offsetting behavior is added to the empirical analysis. According to the rational economic man principle, per capita wage is selected to signify the workers' offsetting behavior. In addition to chemical enterprise behavior and worker behavior, other disturbance factors also influence the performance. In order to control these influences, the chemical gross industry output value index also is included in the empirical model.

[1]The cause of choosing investment in fixed assets as the investment in safety facilities is that the statistical data of investment in safety facilities not listed separately in China, but included in the investment in fixed assets. Another reason is that good working environments and advanced equipment are the safety guarantees for workers.

[2]Peltzman (1975) found that the increase of auto safety equipment did not reduce traffic mortality in the study of automobile safety regulation effect because of the offsetting effect caused by the behavior of drivers. Klick and Stratmann (2003) defined this effect as the Peltzman effect.

This paper uses the annual data[3] over the period 1981-2011 to identify the development trend of China's chemical accidents. Chemical industry refers to the Manufacture of Raw Chemical Materials and Chemical Products according to national economy classifications in China. The number of chemical accidents (1981-2000) is drawn from the book Selected Cases of Major Chemical Accidents, and the number of chemical accidents (2001-2011) is from the AIS (Accident Inquiry System) of the SAWS (State Administration of Work Safety) in China. The data of chemical gross industry output value and investment in fixed assets are calculated from the China Statistical Yearbook (1982-2012). The data about per capita wage is calculated from the China Labor Statistical Yearbook (1991-2012). In order to eliminate price change effects on chemical gross industry output value and investment in fixed assets, we transform the data of chemical gross industry output value and investment in fixed assets to constant price of 1978 by the GDP deflator. To eliminate heteroscedasticity, except for accident incidence, all of the other variables take the form of natural logarithms. These four variables are respectively expressed by S_t, V_t, I_t, W_t.

[3]The data is not reported here to conserve space but is available from the author upon request.

EMPIRICAL RESULTS

Unit Root Test

We begin our empirical analysis by testing for unit roots in the accident incidence, S_t gross industry output value V_t, investment in safety facilities I_t, and per capita wage W_t, because the integrational properties are crucial for the cointegration test and Granger causality test in VECM framework. We apply the conventional augmented Dickey and Fuller (ADF, 1979) test to establish the integrational properties of S_t, V_t, I_t, W_t The calculated t-statistics together with the lag length selected using the SIC (Schwarz Information Criterion), as well as the critical value at 5% for the accident incidence S_t, gross industry output value V_t, investment in safety facilities I_t, and per capita wage W_t series are reported in Table 1.

In Table 1, the calculated t-statistics for the levels of accident incidence S_t, gross industry output value V_t, investment in safety facilities I_t, and per capita wage W_t series are greater than the critical value at the 5% significant level. This implies that we cannot reject the unit root null hypothesis. However, when we convert the accident incidence S_t, gross industry output value V_t, investment in safety facilities I_t, and per capita wage W_t series into first difference and subject the series to the ADF test, the calculated t-statistic for all accident incidence S_t, gross industry output value V_t, investment in safety facilities I_t, and per capita wage W_t is smaller than the critical value at the 5% level. This implies that we can reject the unit root null hypothesis for all series in first difference form. As a result, all variables are integrated of order one. This paves the way for conducting tests for cointegration and Granger causality in a VECM framework later in the paper.

Cointegration Test

In order to examine the long-term relationship among the accident incidence S_t, gross industry output value V_t, investment in safety facilities I_t, and per capita wage W_t, we use the Johansen test method to perform a cointegration test. According to the AIC (Akaike Information Criterion) and SIC (Schwarz Information Criterion), the optimal number of lags for the Johansen cointegration test method is 3. When the trace statistic is greater than the critical value at the 5% significant level, reject the null hypothesis of "no cointegration"; When the trace statistic is smaller than the critical value at the 5% significant level, accept the null hypothesis. The test results are listed inTable 2.

As illustrated in Table 2, we accept the null hypothesis of "at most 3 cointegration relationships existed" among the accident incidence S_t, gross output value V_t, investment in safety I_t, and per capita wage W_t at the 5% significant level. The cointegration equation (two additional cointegration relationships are omitted because they are irrelevant to this paper) is estimated below (the standard errors are in parentheses):

$$EC = S - \underset{(0.12454)}{1.219691V} + \underset{(0.08206)}{0.283725I} + \underset{(0.13543)}{1.435979W} \tag{1}$$

Table 1: ADF unit root test results of S_t, V_t, I_t, W_t

Variable	S_t	ΔS_t	V_t	ΔV_t	I_t	ΔI_t	W_t	ΔW_t
t-statistic	-0.9979	-5.3515	-1.0925	-5.3710	-1.7326	-2.3890	13.0112	-4.3308
[LL]	[0]	[0]	[0]	[0]	[1]	[0]	[0]	[0]
CV	-1.9525	-1.9529	-3.5684	-3.5742	-3.5742	-1.9529	-1.9525	-3.5742

Notes: LL denotes lag length, which is selected using the SIC automatically and CV denotes critical values at the 5% significant level. Δ is the difference operator.

Table 2: Cointegration test results of S_t, V_t, I_t, W_t

Hypothesized No. of CE(s)	Eigenvalue	Trace Statistic	0.05 Critical Value	P
None*	0.9323	146.8033	47.8561	0.0000
At most 1*	0.7701	74.1057	29.7971	0.0000
At most 2*	0.7140	34.4119	15.4947	0.0000
At most 3	0.0224	0.6104	3.8415	0.4346

From Equation (1), in the long term, an increase in the gross output value V_t has a positive effect on accident incidence S_t, and if the gross output value V_t is increased by 1%, the accident incidence S_t will be increased by 1.22%. An increase of investment in safety facilities I_t and per capita wage W_t both have negative effects on accident incidence S_t, and when investment in safety facilities I_t and per capita wage W_t are respectively increased by 1%, the accident incidence S_t will be decreased separately by 0.28% and 1.44%. The negative effect of per capita wage W_t is stronger than the positive effect of gross output value V_t.

VECM

A causality test is often used to analyze the causal relationship among the variables. When there is cointegration relationship among the variables, we can construct a VECM to get the regression equation including the error correction item. The Wald joint test is then used to test the significance of the coefficient both on the variables and the error correction item to judge the causality direction [15] [16] in the VECM framework. The lag length is equal to the lag length for the cointegration test. The general equation of VECM is expressed by:

$$\Delta Y_t = \sum_{i=1}^{p} \Gamma_i \Delta y_{t-i} + \lambda EC_{t-1} + \varepsilon_t$$

$$(2)$$

where $Y_t = [S_t V_t I_t W_t]'$, T_i is the coefficient matrix, reflecting the impact made by short-term change of explaining variables to short-term change of explained variable. $EC_{t=1}$ is the error correction item, reflecting the long-term equilibrium relationship of variables. λ is the coefficient vector of $EC_{t=1}$, reflecting the adjustment velocity from disequilibrium to equilibrium when it deviates from a long-term equilibrium state. ε_t denotes random error vector.

The specific VECM equation[4] in which accident incidence S_t is the explained variable with gross industry output value V_t, investment in safety I_t and per capita wage W_t as the explaining variable is:

[4]The main purpose of this paper is to analyze the impact on accident incidence made by chemical gross industry output value, investment

in safety facilities, and per capita wage. Thus, only the equation in which accident incidence is the explained variable is listed.

$$\Delta S_t = \begin{bmatrix} -0.1976 \\ -0.0158 \\ -0.0134 \\ 0.0712 \end{bmatrix} \begin{bmatrix} \Delta S & \Delta V & \Delta I & \Delta W \end{bmatrix}_{t-1} + \begin{bmatrix} -0.0510 \\ 0.0100 \\ 0.0062 \\ -0.0083 \end{bmatrix} \begin{bmatrix} \Delta S & \Delta V & \Delta I & \Delta W \end{bmatrix}_{t-2}$$

$$+ \begin{bmatrix} -0.2474 \\ -0.0279 \\ -0.0195 \\ 0.1598 \end{bmatrix} \begin{bmatrix} \Delta S & \Delta V & \Delta I & \Delta W \end{bmatrix}_{t-3} - 0.0146 EC_{t-1} - 0.0108.$$

$$\tag{3}$$

where $EC_{t-1} = S_{t-1} - 1.2197 V_{t-1} + 0.2837 I_{t-1} + 1.4360 W_{t-1} - 3.3581.$ From the Equation (3), the coefficient on $EC_{t=1}$ is −0.0146, meaning that the adjustment degree of the disequilibrium for the previous year is the 1.46%.

At the 5% significant level, use the Wald joint test on the equation in which accident incidence S_t is the explained variable, with gross industry output value V_t, investment in safety facilities I_t and per capita wage W_t as explaining variables. The null hypothesis of the Wald joint test is that there is no granger causality between variables. When the probability value of the Wald joint test χ^2 is greater than 0.05, accept the null hypothesis; When the probability value of the Wald joint test χ^2 is smaller than 0.05, rejected the null hypothesis. The Wald joint test results are shown in Table 3.

The statistical significances of each coefficient in Table 3 indicate that at the 5% significant level, all the null hypothesizes (H_0) were rejected. Combined with the VECM equation interpretation, we know that in both the short and long term, the changes of gross industry output value V_t, investment in safety facilities I_t and per capita wage W_t are all the Granger causes of change in accident incidence S_t.

Impulse Response Function

Through impulse response function analysis, the path of influence affected by gross industry output value V_t, investment in safety facilities I_t and per capita wage W_t on current value and future value of accident incidence S_t can be obtained. To avoid variable order affecting the results, we choose the generalized impulse response function to do impulse response function analysis. The impulse response curves are represented in Figure 1 and Figure 2. The horizontal axis denotes the period (here we report only 10 periods; increasing the period does not affect the conclusion), and the vertical axis denotes the response degree.

Figure 1 shows the curve of response of per capita wage W_t to investment in safety facilities I_t impulse. From Figure 1 we can demonstrate that, in the short term, the response of per capita wage W_t to investment in safety facilities I_t impulse is positive. In the long term, the response is negative. For a positive information rush of I_t, the maximum positive response of W_t is achieved in the first period. Starting from the third period, the response turns from positive to negative. The strongest negative response is in the fourth stage, subsequently decreasing gradually. That is, in the short term, an increase in investment in safety facilities will increase the per capita wage of workers. In the long run, the increase investment in safety facilities can reduce per capita wage.

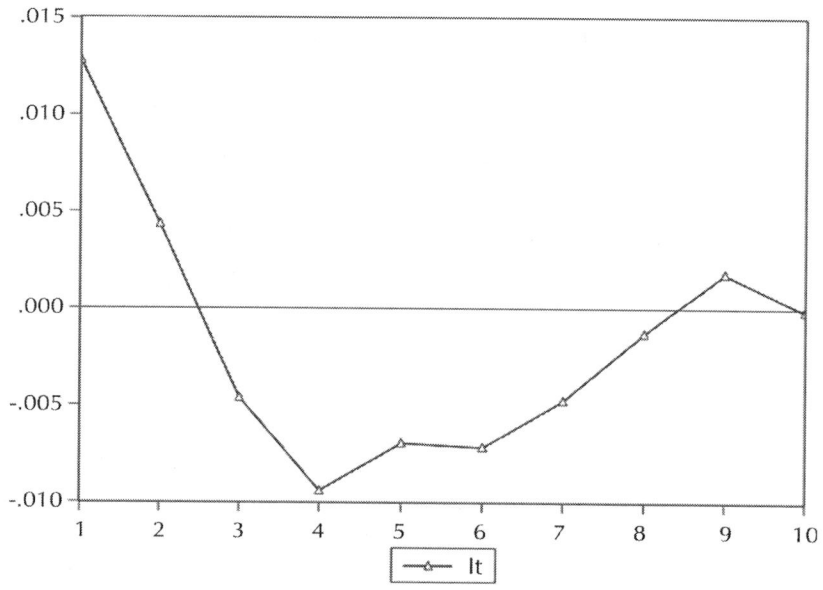

Figure 1: Curve of response of W_t to I_t.

Table 3: The results of the Granger causality test of S_t, V_t, I_t, W_t

Test equation5:	$\Delta S_t = \sum_1^3 [\alpha_{1i}\Delta S_{t-i} + \beta_{1i}\Delta V_{t-i} + \gamma_{1i}\Delta I_{1i} + \delta_{1i}\Delta W_{1i}] + \lambda_1 EC_{t-1} + \varepsilon_1$				
ΔS_t	ΔV_t	ΔI_t	ΔW_t	Joint test	EC test
H_0	$\beta = 0$	$\gamma_{1i} = 0$	$\delta_{1i} = 0$	$\delta_{1i} = \gamma_{1i} = \delta_{1i} = 0$	$\lambda_1 = 0$
χ^2	14.8771	8.3034	20.0818	42.7323	4.03778
P	0.0019	0.0401	0.0002	0.0000	0.0445

Note: H_0 indicates that the row variable doesn't cause the column variable; P is the probability value of Wald joint test χ^2.

[5]The main purpose of this paper is to analyze the impact on accident incidence made by chemical gross industry output value, investment in safety facilities and per capita wage. Therefore, only test the equation in which accident incidence is the explained variable.

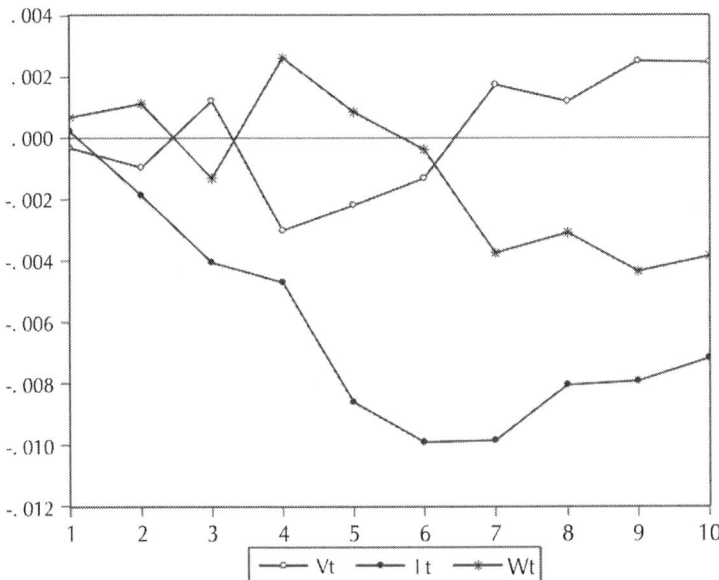

Figure 2: Curves of response of S_t to V_t, I_t, W_t.

Figure 2 shows the curves of response of accident incidence S_t to gross industry output value V_t impulse, investment in safety facilities I_t impulse, and per capita wage W_t impulse. From Figure 2we can deduce: (1) in the short term, the response of accident incidence S_t to gross industry output value V_t is negative. In the long term, the response is positive. For a positive information rush of V_t, the maximum negative response of S_t is achieved in the fourth period. From the seventh period, the response of S_t to V_t becomes positive. These results show that in the short term, the increase in chemical gross industry output value will reduce the accident incidence, but in the long term, the increase will cause the accident incidence to increase. (2) In the short term, the response of accident incidence S_t to investment in safety facilities I_t is positive. In the long term, the response is negative. For a positive information rush of I_t, the maximum positive response of S_t is achieved in the first period. Beginning from the second period, the response becomes negative and continues to strengthen. From the seventh period, the response starts to weaken. That is, in the short term, the increase investment in safety facilities will increase the accident incidence. However, in the long term, the increase in investment in safety facilities will reduce accident incidence. (3) In the short term, the response of accident incidence S_t to per capita wage W_t impulse is positive. In the long term, the response is negative. For a positive information rush of W_t, the maximum positive response of S_t is achieved in the fourth period. Then, the response is negative, and the maximum negative response of S_t is achieved in the ninth period. Namely, in the short term, the increase in per capita wage will cause accident incidence to increase. In the long term, the increase in per capita wage will reduce accident incidence.

From the perspective of the chemical workers' safety awareness, workers tend to generate offsetting behavior, because of workers' behavior in regards to moral hazard. The workers' offsetting behavior is derived from the relatively low knowledge level of chemical workers. The main reason for chemical workers lack knowledge culture is that according to China's family planning policy, each urban family only has one child, and each rural family can have two children. The consequences of the one-child policy are as follows: first, the urban only- child is spoiled and doesn't want to work at the chemical factory in which the working environment is poor; second, the vocational-technical schools in cities designed for chemical plant workers have been forced to close due to lack of students; third, in China, migrant workers[6] have become the current primary labor in high-risk industries such as the chemical industry, especially in private chemical enterprises. Because of migrant workers' education level is relatively low, when judging safety risk, they are more likely to generate offsetting behavior. In the short term, when the enterprises increase investment in safety facilities to provide a safer working environment for workers, the workers will be dependent on enterprise safety precautions excessively, thinking that their working environment is safe, thus reducing their safety awareness[7]. So, workers tend to increase labor efforts to earn higher wages by reducing safety efforts (assume that the worker will allocate his effort between wage effort and safety effort in this paper), leading to an offsetting effect. If the offsetting effect is strong enough, the accident incidence will increase conversely. Therefore, in the short term, with the increase in investment in safety facilities, per capita wage and accident incidence both increase. In the long term, with the improvement of the workers' risk prevention awareness level through strengthening education and training, when increasing investment in safety facilities, workers will be aware of the increase in potential risk, rather than believing that their working environment gets better. Thus, workers will transfer their effort to safety from wages, obtaining a higher safety level. That is, with the workers' offsetting behavior disappearing gradually, an increase in investment in safety facilities will eventually have the effect off reducing accident incidence.

[6]Refers to the agricultural registered permanent residents working at the local township enterprise or urban enterprise.

[7]Viscusi (1979) through study found that when companies improve workers' working conditions and the quality of labor safety, workers' safety efforts will decline and the action level preventing risk will drop.

CASE STUDY

Dalian is located in the northeast coastal chemical industry area of China, and the chemical industry is developing rapidly. After the oil pipeline explosion on July 16, 2010 which shocked the world, the nation and the local government attached great importance to chemical safety. In 2010 and 2011, the government passed a total of 6 laws and regulations about hazardous chemicals (30, 32, 36, 40, 41, 42 in order), two times more than the sum of the past five years. Meanwhile, the Daliangovernment invested 20 million Yuan (investment in safety facilities) to improve the safety level of chemical companies, including the upgrade and reform of enterprise hazardous process automation control, with all the chemical enterprises within the district achieving the process automation control. This 20 million Yuan investment not only achieved the expected effect, but also increased the accident incidence.

On November 25, 2010, a toxic gases spill accident (including carbon oxide, hydrogen sulfide and so on) occurred in Carbon Chemical Co., Ltd. in Dalian, resulting in more than 20 workers, who were doing gymnastics close to the scene, suffering from toxic gas poisoning. The enterprise has sound safety rules and regulations, perfect safety operational procedures and security facilities. The cause of the accident was ash deposits in the gasifier, leading to the stopping of the compressor, and then the system stopped automatically. When the gases in the furnace burned insufficiently, toxic gases emitted directly through the torch. The direct cause of the accident was the workers' over-reliance on the automatic safety chain system. When deviant behavior occurred at the end of the safety chain, there was no human intervention, eventually leading to the gas leakage accident[8].

[8]See the two cases, Dalian Administration Bureau of Safety Working in China (2010-2011).

On August 29, 2011, Dalian Petrochemical Company's refined oil storage tank exploded and caught fire. The incident occurred when the tubing outlet velocity of #875 tank reached 4.34 m/s, due to the floating plate without automatic floating, during the oil delivery operations of refined oil storage tanks. Due to exceeded safety limits, a large amount of static electricity was produced and discharged, igniting the mixture of oil mist, combustible gas and air, which exploded. This was a typical accident caused by violation of rules. When the tubing outlet velocity exceeded the safety limits, the workers didn't control the tubing outlet velocity within safe limits in accordance with the rules, but relied on automatic equipment operation fully, resulting in the storage tank exploding and catching fire.

During the 2010 and 2011, while safety regulation was enhanced, the number[9] of the chemical accidents was 2.7 times more than the total of the previous two years in Dalian. And 62.5% of 2010-2011 accidents belong to the "three violations" category. This indicates that the offsetting behavior is more serious in Dalian. Regulation enhancement did not reduce the accident incidence, but increased the accident incidence, due to workers relying too heavily on security facilities. The main reason is that, several years ago, the original chemical vocational- technical school in Dalian had been forced to close due to insufficient number of students. Migrant workers' safety training is insufficient and safety awareness is weak, which is the reason for serious chemical workers' offsetting behavior in Dalian. The case of Dalian further illustrates workers' offsetting behavior is the main cause of chemical accident increases.

[9]The number of chemical accidents is taken from the China Chemical Safety Association.

CONCLUSIONS

This paper adopts the chemical accident incidence, the chemical gross industry output value, investment in safety facilities of chemical industry, and per capita wage of chemical industry employment as the indexes to empirically analyze investment performance of chemical safety facilities using time series data over the period 1981-2011 by VECM. The empirical results indicate that for China 's chemical industry, in the short term, increasing fails to improve the safety level significantly because of the offsetting behavior of the employees. Over the long term, the offsetting behavior tends to diminish, and the chemical accident incidence can be decreased by increasing the investment performance index. Poor safety awareness among workers is one of the causes of accident incidences. Therefore, making sure to heighten the chemical workers' safety awareness is one of the measures to improve the investment performance of chemical safety facilities in the short term.

Funding

Social Science Fund Project of Liaoning province (L13BJY033); Social Science Fund Project co-sponsored by province and ministry (13JJD790042).

REFERENCES

1.   Sun, K. and Yang, H.M. (2012) Statistical Analysis of Dangerous Chemical Accidents in China. Fire Technology, 48, 331-341. http://dx.doi.org/10.1007/s10694-011-0224-y

2.   Smith, R.S. (1979) The Impact of OSHA Inspection on Manufacturing Injury Rates. Journal of Human Resource, 14, 145-170.

3.   Gray, W.B. and Scholz, J.T. (1993) Does Regulatory Enforcement Work? A Panel Analysis of OSHA Enforcement. Law & Society Review, 27, 177-213.http://dx.doi.org/10.2307/3053754

4.   Lewis-Beck, M.S. and Alford, J.R. (1980) Can Government Regulate Safety? The Coal Mine Example. The American Political Science Review, 74, 745-756.http://dx.doi.org/10.2307/1958155

5.   Carmichael, H.L. (1986) Reputations for Safety: Mark Performance and Policy Remedies. Journal of Labor Economics, 4, 458-472. http://dx.doi.org/10.1086/298106

6.   Weil, D. (1996) If OSHA Is So Bad, Why Is Compliance So Good? The Rand Journal of Economics, 27, 618-640. http://dx.doi.org/10.2307/2555847

7.   Viscusi, K. (1992) Fatal Tradeoffs: Public and Private Responsibilities for Risk. Oxford University Press, New York, 50-171.

8.   Viscusi, K. (1979) The Impact of Occupation Safety and Health Regulation, 1973-1983. Bell Journal of Economics, 10, 117-140. http://dx.doi.org/10.2307/3003322

9.   Keiser, K.R. (1980) The New Regulation of Health and Safety. Political Science, 95, 479-491.http://dx.doi.org/10.2307/2150061

10.   Greenberg, E.S. (1985) Capitalism and the American Political Ideal. M. E. Sharpe, Armonk, 76-80.

11.   Reniers, G.L.L., et al. (2008) A Multiple Shutdown Method for Managing Evacuation in Case of Major Fire Accidents in Chemical Clusters. Journal of Hazardous Materials, 152, 750-756. http://dx.doi.org/10.1016/j.jhazmat.2007.07.040

12.   He, G.Z., et al. (2011) Managing Major Chemical Accidents in China: Towards Effective Risk Information. Journal of

Hazardous Materials, 187, 171-181.http://dx.doi.org/10.1016/j.jhazmat.2011.01.017

13. Zhang, X.M. and Chen, G.H. (2011) Modelin and Algorithm of Domino Effect in Chemical Industrial Parks Using Discrete Isolated Island Method. Safety Science, 49, 463-467.http://dx.doi.org/10.1016/j.ssci.2010.11.002

14. Klick, J. and Stratmann, T. (2003) Offsetting Behavior in the Workplace. Working Paper, George Mason University, Feldstein.

15. Feldstein, M. and Stock, J.H. (1994) The Use of a Monetary Aggregate to Target Nominal GDP. Monetary Policy, University of Chicago Press, Chicago, 7-69.

16. Toda, H.Y. and Phillips, P.C.B. (1993) Vector Autoregressions and Causality. Econometrica, 61, 1367-1393. http://dx.doi.org/10.2307/2951647

Synthesis of Commercial-Scale Tungsten Carbide-Cobalt (WC/Co) Nanocomposite Using Aqueous Solutions of Tungsten (W), Cobalt (Co), and Carbon (C) Precursors

T. Danny Xiao[1], Xinglong Tan[2], Maozhong Yi[2],
Shigao Peng[3], Fangcai Peng[3], Jiangao Yang[4],
and Yu Dai[4]

[1]Inframat Corporation, 151 Progress Drive, Manchester, CT, USA

[2]State Key Laboratory of Powder Metallurgy, Central South University, Changsha, China

[3]Fujian Jinxin Tungsten Co., Ltd., Longyan City, China

[4]Hunan ACME Technology Co., Ltd, ACME Technology Park, Changsha, China

ABSTRACT

This paper reports the chemical synthesis of tungsten carbide/cobalt (WC/Co) nanocomposite powders via a unique chemical processing technique, involving the using of all water soluble solution of W-, Coand C-precursors. In the actual synthesis, large quantities of commercial-scale WC-Co nanocomposite powders are made by an unique combination of converting a molecularly mixed W-, Co-, and C-containing solutions into a complex inorganic polymeric powder precursor, conversion of the inorganic polymeric precursor powder into a W-Co-C-O containing powder intermediates using a belt furnace with temperature at about 500°C - 600°C in an inert atmosphere, followed by carburization in a rotary furnace at temperature less than 1000°C in nitrogen. Liquid phase sintering technique is used to consolidate the WC/Co nanocomposite powder into sintered bulk parts. The sintered parts have excellent hardness in excess of 93 HRA, with WC grains in the order of 200 - 300 nm, while Co phase is uniformly distributed on the grain boundaries of the WC nanoparticles. We also report the presence of cobalt Co precipitates inside tungsten carbide WC nanograins in the composites of the consolidated bulk parts. EDS is used to identify the presence of these precipitates and micro-micro-diffraction technique is employed to determine the nature of these precipitates.

INTRODUCTION

The tungsten carbide-cobalt (WC-Co) composite material, so called cermet or hardmetal, is the most widely used hardfacing materials in industrial cutting or boring tools applications. The study of the tungsten carbide material can be traced back more than hundred years old history. The first discovery of this family of material was a carbon deficient W_2C phase, by a German researcher, Mr. Henri Moissan, in 1896 [1] , by reaction of tungsten oxide with carbon, and the original use of this W_2C material was as an addition to low carbon steel to make high speed tool steels. In 1914, Voigtlander and Lohmann developed a method of manufacturing WC powders by fusing mixtures of tungsten oxide with carbon in a carbon tube furnace [2]. A German company, Osram Studiengesellschaft, invented the first WC/Co composite material, subsequently a first WC/Co patent was issued to this company in 1923 [3] , while the first WC with 6 wt% Co cemented carbide hardmetal appeared in 1926 under the name "Widia" [4] .

Over many decades, using the classic WC manufacturing technique, WC powders were made by the steps of: converting ammonium paratungstate (APT) into blue tungsten ($WO_{2.9}$), reducing of blue oxide to tungsten (W), mixing of tungsten with carbon black (W + C), and reacting of the tungsten/carbon mixture at high temperature to form WC compound. The WC/Co composite powder was then made by simply ball milling the WC powder with metallic Co to form physically mixed composite, and tools or hardmetal cermets were made by shape-forming followed by high temperature sintering. Here, Co serves in two functions, as a binder phase for the tungsten carbide hard phase grains, and to provide toughness for the composite. Composite produced by this method is often associated with large WC grains, in the order of several microns in size, thus mechanical properties, such as hardness and toughness are often limited.

It was found that decreasing the size of the WC phase leads to an increased hardness with simultaneous increased toughness, thus significantly extending the wear life of the consolidated WC/Co hardmetal components. With this expectation in mind, in the past half century since the late sixties, researchers have been investigating many different methods to reduce the size of the tungsten carbide grains from the original several microns in dimensions, down to submicron, ultrafine, and recently down to the nanometer regime.

From 1960 to 1970, both Gortsema at Union Carbide [5] -[7] and Takatsu at Toshiba [8] -[10] had developed methods for manufacturing ultrafine WC that both involved reduction and carburization of tungsten oxide using a mixture gas of either CH_4/H_2 or CO/CO_2, which relied heavily on kinetic rather than thermodynamic control of the carburization step, leading to difficult control of processing parameters in subsequent scale-up productions.

In the late 1970's to 1990's, multi-step carburization processes were invented to produce ultrafine or superfine WC particles. In one of the multi-step processes developed by Miyake et al. [11] -[13] , WO_3 and carbon blacks with selected stoichiometic ratios were mixed together, and the mixture was first carburized in a rotary furnace in an inert atmosphere of nitrogen to form a mixture of WC, W_2C and W at temperatures from 1000°C up to 1120°C. Carbon blacks were then added to the mixture of WC, W_2C and W via a ball milling process, followed by a second carburization step again in rotary furnace at temperatures of 1200°C to 1300°C in a hydrogen atmosphere. The advantage of this process is the easy control of carburization atmosphere during production. However, a drawback was that the high temperature reaction at the second carburization step resulted in partial sintering of WC particles, thus, leading to a required post grinding process to produce final product with a particle size of 0.1 to 0.2 microns. This process has been adopted today by Tokyo Tungsten in the manufacturing of ultrafine or superfine WC powders. Using a similar method, Xiao et al. at Inframat Corp. in the 2000's [14] , had produced nanometer size WC powders via first a full carburization of high energy ball milled WO_3 and carbon blacks into WC powders with excess carbon content in a nitrogen atmosphere with temperatures of 1000°C - 1150°C, followed by subsequent fine carbon adjustment to stoichiometric WC in hydrogen/hydrocarbon mixture at temperatures less than 1000°C. This approach, although had been proven to be useful for thermal sprayed coating powder feedstock in surface engineering and other industrial tool applications [15] [16] , but it will required further research to eliminate the presence of minor larger than 1 micron carbide particles for which is the requirement of micro-drills for electronic cutting tool applications.

Another multi-step carburization process has been developed by Zuker [17] . In this method, WC was synthesized from a tungsten precursor compound by heating the precursor compound to a first temperature at least 450°C in a reducing gas composition to form an intermediate tungsten product, and then carburizing the intermediate tungsten product in a second furnace at temperature above 750°C under a carburization condition in a hydrocarbon gas environment. Later, Dow Chemical had modified this process to produce commercial quantities of 0.1 - 0.2 micron size WC powders.

Kear et al., in the late 1980's and early 1990's, had successfully made WC/Co nanocomposite powders via a spray conversion process [18] [19] . In this technique, soluble tungsten, ammonium metatungstate (AMT), and cobalt, cobalt nitrate, acetate or oxalate, were dissolved in water to make aqueous solution, followed by a spray conversion process to make W-Co-O containing precursor powder, with subsequent reduction, and carburization to convert it into WC/Co at temperatures of 700°C to 900°C, in a CO/CO_2 gaseous environment in a fluidized bed. This process had been the classic method of making WC/Co nanocomposite, and had opened the window opportunity for obtaining nanocomposite materials with significantly improved structural and mechanical properties [20] [21] . This process, however, although had been proven to be successful in making small pilot-plant quantities of WC/Co nanocomposite powders, due to the intrinsic characteristic of obtaining carbides via a gassolid reaction mechanism for carburization in a fludized-bed reactor which heavily depended on reaction kinetics, scale-up of this process into commercial production becomes a real challenge.

Followed the initial invention of Kear, et al., many research groups, including Chinese researchers at Zhuzhou Cemented Carbide [22] , and Shao's group at Huazhong Science & Technology University of China [23] , had performed extensive research in making WC/ Co nanocomposites. In these years of research, efforts were made to modify the process so that uniform and large quantities WC/Co nanocomposite can be made for significantly improved hardness and toughness in hardmetal applications.

This paper reports a new chemical synthesis process of making WC/Co nanocomposite material. Distinct difference from Kear's processing method, however, is that this research involves the processing of WC/Co nanocomposite using all water soluble solution of W-, Coand C-precursors. In this process, an aqueous solution of W-, Coand C-containing chemical precursors is obtained via dissolving of respective water-soluble chemicals with selected stoichiometic ratio. This precursor solution is then spray converted into powders, followed by eliminating residual moistures, and organic ligands such as N-H and C-H groups to form a W-Co-C-O intermediate powder with controlled stoichiometry in a belt furnace under inert atmosphere either in argon or nitrogen. The last step is to convert this intermediate powder into WC/Co nanocomposite by a carburization process in a rotary furnace to obtain nanostructured WC/Co powders at elevated temperatures under nitrogen. Similar to Kear's spray conversion process, this process mixes the W and Co sources in a molecular level. Additional advantages are that the C molecules are also uniformly mixed with the W and Co elements at the molecular level, thus ensure the uniformity of chemical reactions to form WC/Co composite in the carburization step under thermodynamic control, which warrants the uniform distribution of carbon atoms prior to the carburization step, and the ability of scale-up in a production environment for making uniform and high quality WC/Co nanocomposite powders.

Consolidations of the bulk ingots are performed using a conventional press-sintering technique at the liquid phase sintering temperature. Thin specimens are then prepared from the consolidated samples for microstructural analysis. Results of both synthesized powder and consolidated nanostructured WC/Co samples are reported here. Due to the molecular level mixing of W-, Co, and C-ingredient at the starting precursor, we have observed special microstructures of the consolidated WC/6 Co in this study, the entrapment of Co nanoparticulates inside WC nanograins. The special microstructures may provide a clue for reaction mechanism of how WC/Co nanocomposite is formed during synthesis.

This paper will deal with the report of making WC/Co nanocomposite in a large quantity manner using a spray conversion technique in combination with belt and rotary furnaces, and some preliminary results of processing these powders into bulk parts. The mechanism of chemical reactions in each steps of the synthesis and the detailed studies of Co nanoparticulate dispersions presented inside WC nanograins will be reported in subsequent publications elsewhere.

EXPERIMENTAL

Synthesis and Fabrication

The processing procedures for making WC/Co nanocomposite powders are shown in sequential steps schematically shown in Figure 1. It involves: 1) preparation of precursor solutions by dissolving W-, Co-, and C-containing precursors into DD water with selected stoichiometric ratios; 2) spray drying of the precursor solution to form precursor powders; 3) conversion of the precursor powder in a belt furnace to form W-Co-C-O containing pre-composite powders; and 4) dissociation of oxygen and carburization of the pre-composite powder into final WC/Co nanocomposite powder in a rotary furnace.

Nanostrucrued 94 WC/6 Co and 88 WC/12 Co powder is made by dissolving selected ratio of AMT, cobalt nitrate, and water soluble carbon source such as polyvinyl alcohol (PVA), corn starch or glucose. The selected ratio of W to C being 1:4 by atomic ratio, with Co to WC ratio being 6 and 12 weight percent for the 94WC/6Co and 88 WC/12 Co composite powders, respectively. The solid precursor chemicals, tungsten source of AMT, cobalt source of nitrate, and carbon source of PVA or starch are then dissolved into DI water with the solid to water ratio being 2:1 by weight. The solution is under vigorous mechanical stirring until a clear liquid without obvious precipitates can be seen by naked eyes. Next, the dissolved aqueous solution is fed into an industrial spray drier, Model No. ACEM 100K (Denli Corp., P.R. China), to form a powder like precursor. This spray dryer has the capability of evaporating 100 kg water per hour. Next, the spray dried precursor powder is fed into a belt furnace, Model No. GTH 1500/100-14 (Denli Corp., P.R. China), with a 200 kg per hour feeding capacity, under a stream of argon or nitrogen atmosphere, to keep a positive pressure inside the furnace to prevent possible moisture re-precipitation and powder oxidation during conversion. This operation is desired to remove residual moisture and crystalline water, dissociation of ammonia N-H groups into hydrogen and nitrogen, dissociation of C-H groups into carbon source and hydrogen, and finally convert the precursor ingredients into a W-Co-C-O complex oxide structure, here referring to a pre-composite powder. The temperature of this process is between 500°C to 600°C with a scheduled temperature profile. Subsequently, the obtained pre-composite powder is then converted to a WC/Co nanocomposite powder, with particle size being 30 to 100 nanometers in a rotary furnace (Model No. HHC- 400 × 7600, Denli Corp.) under nitrogen. If in some case, extra free carbon is presented in the powder, the powder is then again fed into a rotary furnace under stream of hydrogen to remove extra carbon species.

Fully carburized nanocomposite powders are then consolidated into bulk ingots using conventional high temperature liquid phase sintering technique. To prevent abnormal grain growth, ultrafine 0.3 wt% Cr_3C_2 and 0.2 wt% VC particles are added to the WC/Co powder by a ball milling process, followed by adding wax into the mixture. The powder mixture is pressed into green-shaped pellets, followed by sintering of these pellets in a high temperature vacuum furnace with temperature of about 1380°C in ~ 1 h to form sintered ingots with near 100% density. The sintering steps include 550°C dewaxing at about 0.5 h, ~1100°C pre-sintering in ~1 h, and 1380°C sintering in ~1 h.

Characterization

Powder Samples

Powders obtained from both spray dried, belt furnace converted, and rotary furnace carburized are fully characterized. Morphological examination of the as-synthesized powders is obtained by SEM. Particle size is performed using SEM and laser scattering technique instrument (Malvern AWM200). Respective chemistry and phase composition of the nanocomposite is performed using powder diffraction technique (XRD) (Bruker AXS Model No D2 Phaser A26-X1-A). Average particle sizes of WC crystallites are determined using XRD peak broadening analysis technique according to Scherrer equation. The calculated crystallite size is also confirmed using transmission electron microscopy technique.

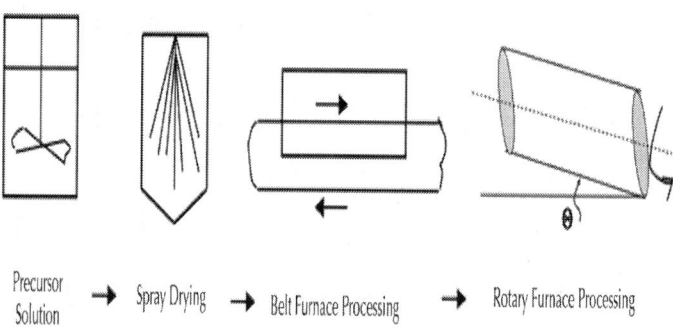

Figure 1: Tungsten carbide/cobalt WC/Co nanocomposite processing flow-chart.

Leco instruments is used to determine both amount of total carbon (carbon detector model No. 619-600-400) and free carbon, and oxygen (oxygen detector Model No. 631-800-100). The Co elemental analyses are performed using wet chemical techniques. Powder surface area is determined using a Quantachrome (model no. 2SI-9 Quadra Sorb SI) surface analyzer. Microstructures of the rotary furnace processed WC/Co nanocomposite powders are also studied using transmission electron microscopy techniques.

Consolidated Bulk Samples

Thin sections of WC/Co specimens are prepared by first electron discharge machining of these sintered pellets into 3 mm diameter with less than 0.1 mm thick discs, followed by ion beam thinning at ~5 kV until perforated. Note, a similar procedure had been used by Mohan and Strutt in the 1990s in preparation of nanostructured WC/Co thin specimens for TEM examination [20]. Characterizations of these specimens are performed by SEM, TEM, micro-probe, and EDAX. To observe these nanoparticulate dispersions inside WC nanograins, a special technique using high resolution TEM is performed with JEOL-100 CX microscopy. The specimens on which the analytical work is done were mounted on a cryo-stage and cooled to below −150°C to minimize thermal effect. Micro-micro-diffractions (μμd) are performed by focusing the electron beam into approximately 200 Å in spot size. Micro-probe with EDAX capability are also performed to identify chemical composition of these precipitated nanoparticulate dispersions inside the WC nanograins.

EXPERIMENTAL RESULTS AND DISCUSSIONS

Spray Converted Powder General Features

The as-spray dried powder has a dark red-color as shown in Figure 2(a). Typical SEM micrographs of the spray drying powder are shown in Figures 2(b)-(d). The general morphology of the as-spray dried powder is spherical in nature (see Figure 1(b)), having smooth featureless surface structure (Figure 1(c)). Broken spheres are also shown in Figure 1(d), which revealing hollow spheres. The measured apparent density of the spray dried powders is ~0.85 g/cc. Powder residual moistures are measured by taking 100 grams of powder, placed into a furnace of ~100°C temperature drying in 4 hours. The obtained residual moistures are between 1 wt% to 3 wt% depending on the humidity and temperature of the spraying time.

Figure 2: Spray dried powders (a) photo revealing a dark red color; (b) low magnification SEM showing the general spherical morphology; (c) increased magnification; and (d) same magnification of (c) of some broken spheres showing that the powder has a hollow shell spherical morphology.

The spray drying process is used to convert the precursor solution mixture of W-, Co-, and C-containing compounds into precursor powder. In this process, liquid precursors are atomized into fine droplets via a rotary atomizer in the presence of hot air at temperature to above 200°C. In this process, droplet surfaces are rapidly heated while liquid solvent (water) are being evaporated along with the shrinkage of the droplet size. Once the majority solvent is removed by evaporation, solid powders are formed. The size and morphology of the powder for a given dimension of the droplets depends on the rate of the solvent evaporation process as well as the temperature distribution in the droplet traveling path within the spray dryer. In other words, if the rate of the evaporation is larger than the rate of droplet shrinkage, hard skins will be most likely to form first on the outer surface while air and residual moisture are trapped inside the droplets during drying process, resulting in hollow shell spheres. Thus, the morphology or the size of the hole (or hollow) and density of the obtained powders is a function of temperature, liquid concentration, droplet size, and spray drying temperature.

The obtained powder, if it is heated to above 200°C in air, will form into a sticky melting gel with rapid volume expansion to about 10 times of its original dimension, which after cooling to room temperature will form a low density foam structure solid. It seems that the spray dried powder had some degree of reaction between 3 different kinds of precursors, namely, ammonium metatungstate (AMT), cobalt nitrate, and the C-H groups of carbon source. The resulting material is most likely being a metalorganic polymer with cross-linked complex N-H and C-H groups in the structure. The exact structure of this powder is not fully understood at this time, and further research is being conducted and results will be published later.

Before converting the as-spray dried powder into WC/Co nanocomposite, we had conducted TGA/DSC experiments to further investigate its behavior at elevated temperatures. In the TGA/DSC studies, 31.4 miligrams of the 6% Co spray dried precursor powders are placed into a platinum crucible, under a stream of nitrogen, heated from room temperature up to 1100°C with heating rate of 10°C per minute. Results of the TGA/DSC curve are shown in Figure 3.

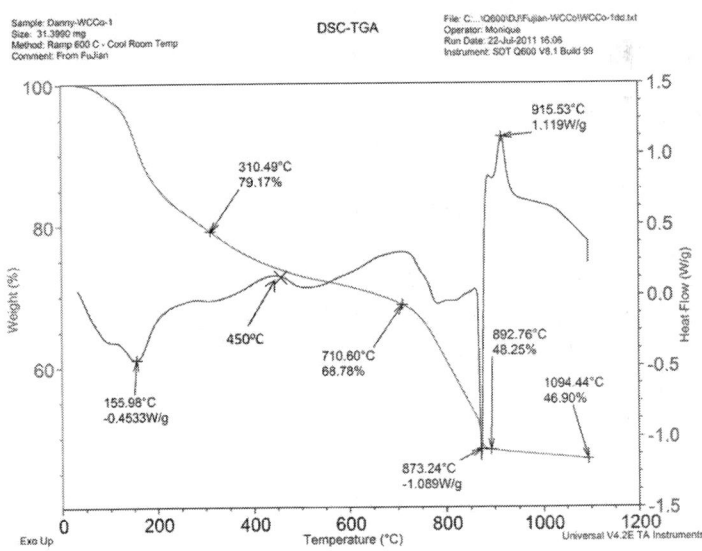

Figure 3: Results of the TGA/DSC curve showing weight and heat flow characteristics; with temperature from room to 1100°C, heating rate of 10°C/minute, 31.4 miligrams powder.

The TGA curve indicated a total loss of 51.75% of weight at temperature of ~900°C, corresponding to a residual weight of 48.25%. Theoretical calculation of the input ingredient is 48.07% assuming all the precursors are converted to WC/6 Co in the formula, assuming 100% chemical reaction. In reality, there is always residual oxygen presented in the WC/Co composite, which is ~0.1% - 0.2%. As observed in the TGA curve, reaction mechanisms for the formation of WC/Co from this ingredient system may follow the reaction sequence shown in Figure 4.

The corresponding major DSC heat flow peaks are at temperatures of 156°C, 310°C and 450°C for removal of crystal water, C-H dissociation /formation of carbon blacks, and dissociation N-H gropus and formation of the W-Co-C-O system; while the temperature of 710°C, 873°C and 915°C could be most likely corresponding to the dissociation of W-Co-O, followed by immediate formation of the WC phase.

Followed the TGA/DSC curve, we understand that the conversion of the precursor powder into nanocomposite powder is quite complex. That is the precursor has to follow a reaction sequence including: 1) residual moisture and crystal water removal; 2) conversion of C-H groups into a solid carbon source; 3) dissociation of N-H groups; and 4) dissociation of the W-Co-O complex compound followed by immediate reaction of the W and C to form the WC/Co nanocomposite. To reduce the complication in the synthesis process, we therefore added another process step, namely belt furnace conversion that are described in the following section.

Belt Furnace Converted Powders

The belt furnace processing here serves the purpose of: 1) removal of residual moisture resulting from spray drying process and crystal water that are inherited from the chemical precursors of AMT, cobalt nitrate, and the aqueous carbon source; 2) conversion of the C-H from the carbon source into carbon, with minimal loss of the carbon due to reaction of carbon atoms with surrounding oxygen and hydrogen to form gaseous species including CO and steam (hot H_2O molecules), or some other gaseous species; and finally 3) removal of the N-H groups, thus forming an intermediate pre-composite of W-Co-O-C system, probably in the form of cobalt tungstate $CoWO_4/WO_3$ mixture, and fine dispersions of carbon black.

The belt furnace converted powder has a black color, as shown in Figure 5(a). Typical SEM micrographs of the belt furnace converted powders are shown in Figure 5(b), and Figure 5(c). The powder also has a spherical morphology, with many broken shell structure as shown in Figure 5(b). In contrast to the as-spray dried powder, the belt furnace converted powder reveals a rough surface structure, and higher magnification SEM is shown in Figure 5(c), revealing a fine structure morphology, with particles are in the submicron or nanometer range.

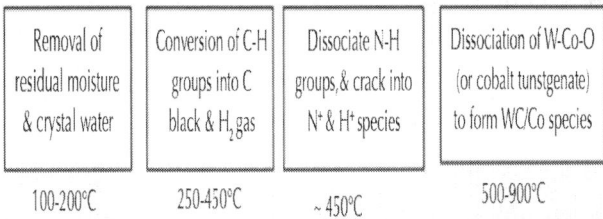

Removal of residual moisture & crystal water	Conversion of C-H groups into C black & H_2 gas	Dissociate N-H groups,& crack into N^+ & H^+ species	Dissociation of W-Co-O (or cobalt tunstgenate) to form WC/Co species
100-200°C	250-450°C	~450°C	500-900°C

Figure 4: Possible reaction sequence with data derived from the TGA/DSC curve.

Figure 5: Belt furnace converted powders showing (a) photo (black powder); (b) low magnification SEM micrograph, the general spherical shell morphology with areas of broken shells; and (c) increased magnification, particles are in the submicron and nanometer regime.

The measured powder apparent density can be controlled to a range of between 1.2 g/cc to 1.8 g/cc, depending on the cobalt content and experimental parameters. These parameters include cobalt composition, gas flow rate, as well as heating profile. Generally speaking, with 6% Co, the obtained apparent density is usually at the lower regime to about 1.2 g/cc or slightly higher, while the 12% Co will have apparent density at about 1.8 g/cc or slightly lower. With a given composition, different heat temperature profile and nitrogen flow rate will result in slightly different powder density. Therefore, in the actual processing, we try to keep the belt furnace at a slightly positive nitrogen flow, with temperature profile of the belt furnace follows the reaction temperature for water remove, C-H group dissociation and conversion as well as N-H group removal temperature profile as indicated in the TGA/DSC data explanation as shown in Figure 4, and the furnace has to create an environment such that as soon as the by-products are created they are immediately removed in the furnace.

As mentioned earlier, the spray dried precursor powder when heated at elevated temperatures in air on a hot plate will result in low density solid with volume expansion as large as 10 times of its original dimension. Similar phenomena had observed in the belt furnace process step in our earlier experiments. Figure 6(a) is photo taken from the belt furnace processed material when the belt furnace processing is not will controlled, and Figure 6(b) is a sketch of the foamed structure.

During processing, a large volume of gas or vapor will be escaped to the top surface powders when powders entered into the high temperatures zone. The initial vapor will largely be hot steam formed by residual moisture and the dissociation of crystal water in the powder. If they are not removed rapidly from the top surface, the hot steam will then cause top surface layer being molten or wet and further drying of the melted surface will result in the formation of a hard cake-like structure or thin hard shell that blocks the vapor escaping path or continuous vapor traveling from below to the top surface. In other words, when the gas expansion rate is much higher than its removal rate, foamed structure will be resulted. This probably explains in many cases when gas delivery rate and the heating profile in the process is not adjusted properly, foamed structures of the belt furnace processed material often resulted as shown in Figure 6. Thus, in later experiments, we have re-adjusted the processing parameters to avoid foam formation. We have found that the best processing temperature is ~500°C - 600°C, with nitrogen flow rate of about ~100 m³/h for powder feeding rate of ~150 kg/hour in this furnace.

WC/Co Nanocomposite Obtained from Rotary Furnace

General Morphology, Density, and Elemental Analysis

The as-synthesized WC/Co powder has a dark-grey color (see Figure 7(a)). Examination via SEM shows that the as-synthesized powder particles are microspheres with hollow-shell microstructure. The powder has apparent density ranging from 1.7 g/cc to 2.7 g/cc depending on the degree of the wall thickness of the hollow shells and experimental conditions that performed in the belt furnace processing. As shown by SEM in Figure 7(b) and Figure 7(c), the hollow-spheres have highly porous microstructures, there are also some highly porous solid spheres microstructure existed in the SEM. Higher magnification using field emission SEM (FESEM) at 30,000× revealed that the composite has facet or plate-like grains (see Figure 7(d)), with particle size ranges from 30 to 100 nanometers in diameter. Energy dispersion analysis is shown in Figure 8, reveals the presence of W, Co and C elements, as well as a small amount of O elements. The presence of O must be due to the surface oxidation of either Co or the WC nanoparticles. It should be noted that this overall morphology of the WC/Co nanocomposite synthesized using the present process via belt furnace combined with rotary furnace carburization in inert atmosphere is very similar to the WC/Co nanocomposite powder obtained via the spray conversion technique using a fluidized-bed reactor in CO/CO_2 solid-gas carburization environment [18] [19] , which also revealed highly porous hollow shell spherical particles.

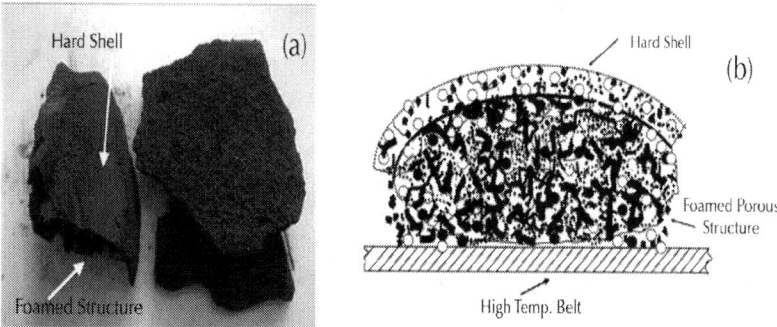

Figure 6: Belt furnace processed material in the case when processing parameters are not well controlled showing (a) photo (left) of the foamed structure, and (b) schematic sketch of the foamed porous structure where at the top is a hard shell, and below the hard shell is a high porosity foamed structure.

Figure 7: Micrographs of the as-synthesized WC/6 Co powder, SEM showing spherical hollow shell particles morphology of (a) photo (dark grey color); (b) general morphology 200× and (c) higher magnification of (b) 2000×, and (d) higher magnification of (c) using FESEM at 30,000× showing the WC/Co nanoparticles having particle size between 50 to 100 nm.

Structural and Phase Determination

Figure 9 is a set of XRD patterns summarizes the structural evolution of the powder undergoing different sequential processing steps, starting from spraying, to belt furnace conversion, and finally to the rotary furnace carburization for the 12% Co nanocomposite composition. The as-spray dried powder reveals an amorphous form in the XRD spectra, while the XRD spectra of the 600°C temperature in nitrogen belt furnace converted pre-composite powder is quite complex, which contains two major phases of cobalt tungstate, $CoWO_4$, and tungsten trioxide, WO_3, as well as a trace amount of cobalt oxide, Co_2O_3. When the belt furnace processing temperature is raised to above 700°C under nitrogen, the obtained XRD pattern reveals a reduced oxide phase of tungsten, namely tungsten dioxide WO_2, and a cobalt tungstate phase $CoWO_4$. The XRD spectra of the rotary furnace carburized powder under nitrogen atmosphere at ~900°C - 950°C shows clean spectrum of major tungsten carbide phase peaks of <001>, <100>, and <101> at 2Q angles of ~31.5°, 35.6°, and 48.3°, respectively, along with a minor phase fcc cobalt with peak <002> at 2Q angle of ~44°. Similar trends in the XRD analysis have been obtained for the 6% Co tungsten carbide/cobalt nanocomposite synthesis.

The average grain size calculated from the XRD peaks using the Scherrer equation is ~50 nm for the powder carburized at ~950°C temperature, while the 900°C temperature carburized powder has WC grain size being ~30 nm and the 1000°C carburized WC grain size is about 80 nm.

Chemical Composition

The total carbon composition of the nanocomposite powder after rotary furnace process is usually at the level of ~5.4 ± 0.05 wt% for 88 WC/12 Co and ~5.8 ± 0.05 wt% for the 94 WC/6 Co, and free carbon content are controlled to at a level of <0.1 wt%. We have found that the carbon control in the rotary furnace process is quite complex, requires a proper ratio of nitrogen flow rate in respect to the powder feeding rate. In our process, we usually set the nitrogen flow rate at a level ~10 - 20 m^3 per hour when we feed the powder at ~25 kg/hour. In many case, if the required carbon did not fall into the required range or slightly off the stoichiometric ratio, namely ~5.4 ± 0.05 wt% for 12 Co and ~5.8 ± 0.05 wt% for 6% Co, a repeated rotary furnace treatment is required to fine tuning the final carbon composition under a mixture gas of methane/hydrogen or carbon monoxide/carbon dioxide at temperatures of above 900°C, and the ratio of respective gas mixture in the process actually depends on the total carbon content presented in the composite. In other words, the fine carbon tuning rotary furnace environment can either be in a carburizing or reducing environment depending on the total carbon required being added or removed in the composite.

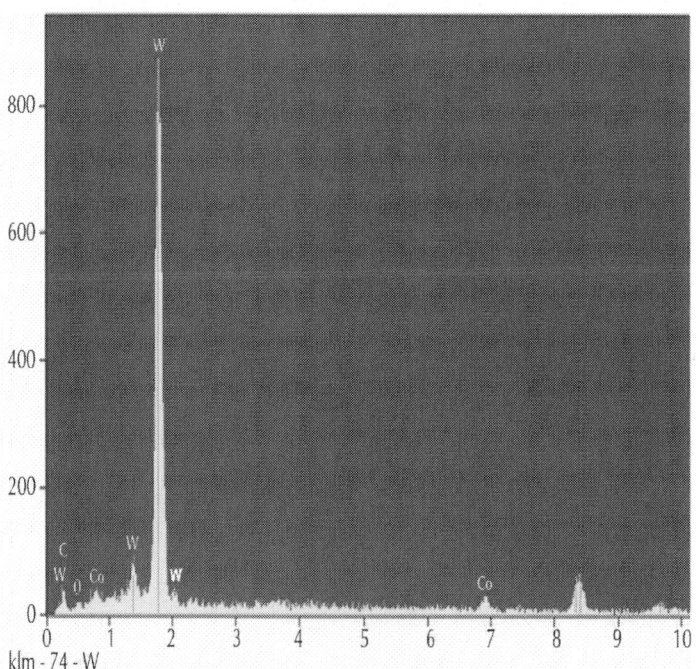

Figure 8: EDS spectrum for showing overall chemistry of the WC/6 Co powder revealing W, C, and Co elements.

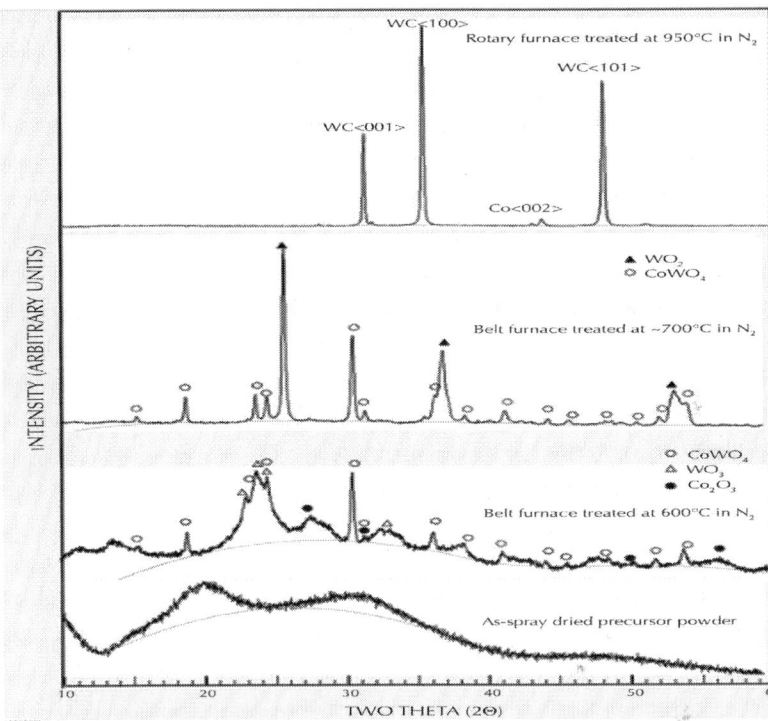

Figure 9: XRD spectra showing (a) the as-spray dried precursor powder exhibiting an amorphous structure; (b) belt furnace processed at ~600°C exhibiting 3 oxide phases of cobalt oxide Co_2O_3, tungsten trioxide WO_3, and cobalt tungstate $CoWO_4$; (c) belt furnace processed at ~700°C exhibiting 2 oxide phases of tungsten dioxide WO_2 and cobalt tungstate $CoWO_4$; and (d) rotary furnace processed at ~950°C exhibiting a composite structure of tungsten carbide phase WC and an fcc cobalt phase Co.

We initially used a typical wet chemical technique to determine the cobalt composition used in carbide industry for WC/Co composite powder analysis. In this technique, 0.15 - 0.2 g of sample is weighed and placed into a 300 ml beaker, 10 ml of solution mixture ($HCl:HNO_3:H_2O$ = 1:3:1) is added to the beaker, and heated to temperature of ~80°C for 2 min; cool down the solution to room temperature and residuals of WC particles are being filtered out using filter papers; dilute the solution using DD water to about 200ml in volume; add 1.5 g $C_6H_{12}N_4$ (hexamethylenetetramine, methenamine) buffer solution, 2 - 3 drops of $C_{23}H_{25}CN_2$ (malachite green oxalate) background solution, and 2 - 3 drops of $C_{31}H_{32}N_2O_{13}S$ (xylenol orange disodium salt) color indicator solution; finally the EDTA (ethylenediaminetetracetic acid) standard solution is slowly dropped into the beaker until the violet color of the solution disappeared. Record the volume of EDTA solution being used, thus the cobalt composition is determined.

The results of the cobalt content are about 5.0 wt% - 5.4 wt% for the 94 WC/6 Co, and 11.0 wt% - 11.4 wt% for the 88 WC/12 Co materials as indicated in Table1 We understand the carbide industry requires that the standard deviation for cobalt is ±0.5 wt% of the specified amount, that is for 88 WC/12 Co material, the Co% should be >11.5 wt%; while for the 94 WC/6 Co material, the Co% should be >5.5 wt%.

To understand this deviation, we explored another detection method, namely the dissolution technique, to determine the Co composition, where the carbide industry normally uses it to determine the cobalt content for sintered WC/Co alloys. In this technique, 0.15 - 0.2 g of sample is weighed and placed into a 300 ml beaker, 10 ml of H_2SO_4 and 5 g of $(NH_4)_2SO_4$ is added to the beaker, and heated to temperature of ~80°C for at least 0.5 h until the solid is fully dissolved; cool the solution to room temperature and adjust pH to about 5.5 - 6.5; add 1.5 g $C_6H_{12}N_4$ (hexamethylenetetramine, methenamine) buffer solution, 2 - 3 drops of $C_{23}H_{25}CN_2$ (malachite green oxalate) background solution, and 2 - 3 drops of $C_{31}H_{32}N_2O_{13}S$ (xylenol orange disodium salt) color indicator solution; finally the EDTA (ethylenediaminetetracetic acid) standard solution is slowly dropped into the beaker until the violet color of the solution disappeared. Record the volume of EDTA solution being used, thus the cobalt composition is finally determined. The results of the cobalt composition are about 5.93 wt% - 6.00 wt% and 11.75 wt% - 12.00 wt% for the 94 WC/6 Co and the 88 WC/12 Co materials, respectively as indicated in Table1

Microstructural Analysis

In the TEM studies, powders are dispersed in ethanol using ultrasonic agitation. Samples are then taken out by dipping carbon coated grids of 200 mesh size into the WC/Co suspension. Both bright-field and dark field images as well as electron diffraction patterns are obtained. TEM bright-field image of the WC/Co nanocomposite powder had showed that the majority of WC grains are well developed facet structure. Image analysis obtained from TEM micrographs reveals that the average WC particle size of ~50 nm, consistent with the particle size obtained. Electron diffraction reveals that existence of both hexagonal WC phase and fcc cobalt phase. Detailed TEM results along with the composite formation mechanisms in the chemical synthesis will be published elsewhere.

Table 1: Experimental results for cobalt composition

Sample ID	Theoretical % Co	Experimental % Co	
		Filtration	Dissolution
88 WC/12 Co	12 wt%	10.9 - 11.3 wt%	11.75 - 12.0 wt%
94 WC/6 Co	6 wt%	5.0 - 5.4 wt%	5.93 - 6.0 wt%

Characteristics of the Consolidated Samples

Although there is a molecular mixing among the ingredients of W, C, and Co during synthesis process, the obtained WC/Co nanocomposite powder has a high porosity hollow shell structure, as shown in Figure 10(a) SEM micrograph of particle morphology in cross-section view, and Figure 10(b) of the schematic illustration of hollow spherical shell cross-section. These hollow shell spheres have particle size ranging from few microns to ~100 microns, where each sphere is an assemblage of millions of WC nanograined particle of ~50 nanometer in size uniformly distributed in a cobalt matrix. Due to a relatively high temperature exposure at above 900°C during rotary furnace carburization, WC nanograins and Co phase have experience some degree of alloying effect due to cobalt diffusion and partial sintering, resulting in a relatively hard hollow shell structure. These hollow shells require extensive milling before they can be pressed into high quality green-body for subsequent sintering process. Both conventional milling and high energy milling process are used to break these shells into particles less than 0.5 microns. In conventional ball milling, 72 to

96 hours are used, while 6 - 8 hours are required when high energy milling process is used. To avoid grain growth during high temperature sintering, small particles of (less than 1 micron) 0.3 wt% Cr_3C_2 and 0.2 wt% VC are added to the powder, followed by milling.

Figure 11 is an SEM reveals typical cross-sectional view of sintered parts for WC/6 Co nanocomposite, illustrating a two phase structure having WC facet grains with crystallite size in the neighborhood of 200 - 300 nm, where the Co material is uniformly distributed on the WC grain boundaries. Preliminary mechanical property investigation reveals that the sintered 12 Co parts are: 93 HRA, 14.2 g/cc density, ~36.1 kA/m coercivity, while the sintered 6 Co parts are, >94.5 HRA, 14.7 g/cc density, and ~44.1 kA/m coercivity.

Transmission electron microscopy technique both in conventional mode and atomic resolution mode are used to investigate the detailed microstructures of the sintered parts. TEM image in bright-field mode for the 6 wt% Co nanocomposite is shown in Figure 12(a), reveals that the consolidated WC grains are being 200 - 300 nanometers in size, with well-developed facet WC crystals. It should be emphasized here that spots features or precipitates are observed within these WC grains (see Figure 12(a) and Figure 12(b)). Micro-micro-diffractions using a ~200 Å spot size are performed focused on these precipitates in Figure 12(b)showing major diffraction spots along with satellite (weak) spots as showing in Figure 12(c) and Figure 12(d). Preliminary studies have identified these major diffraction (strong) spots as the hexagonal WC phase with a = 2.906 Å, c = 0.2838 Å, and the satellite diffraction spots as the fcc Co phase with a = 3.544 Å. Detailed microstructural analysis using high resolution TEM technique is in progress to fully understand these nano-precipitates in relationship with the WC nanoparticles, and the results will be published elsewhere.

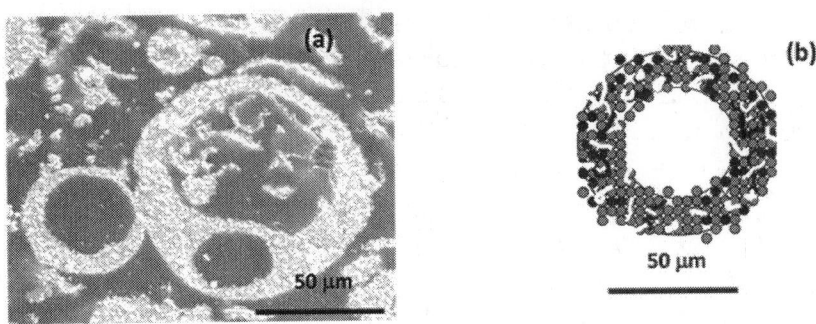

Figure 10: Cross-section of the WC/Co nanocomposite powder showing hollow shelled spherical morphology (a) SEM micrograph; and (b) schematic illustration of the powder cross-section.

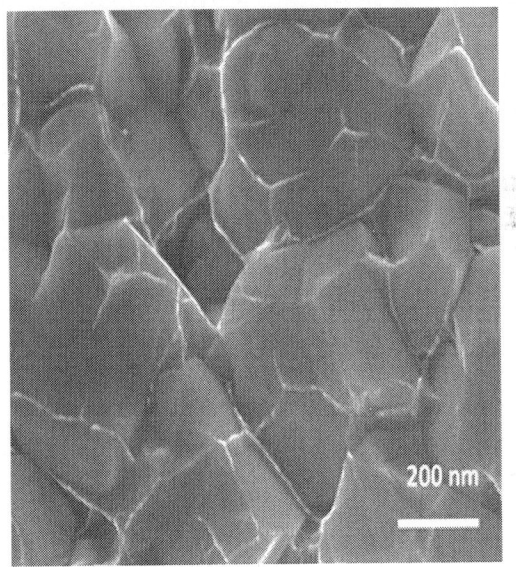

Figure 11: SEM micrograph of consolidated WC/Co nanocomposite revealing WC facet grains along with liquid phase Co uniformly distributed on the WC grain boundaries for 6 wt% Co (40,000× cross-section view of fractured surface).

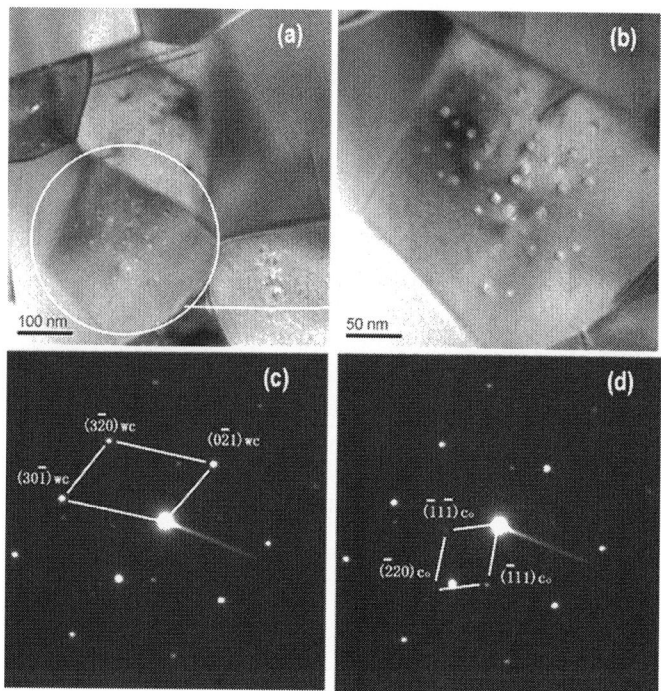

Figure 12: Transmission electron microscopy (TEM) studies showing (a) brightfield image of the liquid phase consolidated parts for WC/6 Co revealing ~200 nm WC facet grains; (b) higher magnification of (a) revealing a secondary phase <10 nm Co nanoparticles precipitated out within the WC nanoparticles; (c) micro-micro diffraction (μμd) of grain in (b) focusing on these precipitates revealing a major diffraction spots identified as the WC phase; and (d) identification of the μμd satellite spots as the Co phase.

It should be emphasized that similar Co-precipitates had been discovered by Mohan and Strutt in their studies of consolidated WC/Co nanocomposites obtained by the spray conversion process [20] . In this study, as described in previous section, we have also noticed that when the chemical analysis of Co was performed by a filtration method, where only the Co phase are dissolved but the WC phase remained solid, there is a ~0.5 wt% shortage in Co content compared to the theoretical values, most likely due to the fact that these Co precipitates residing inside the WC grains cannot be dissolved in the analysis contributing to the chemical deviation. However, when WC and the Co phase are fully dissolved during chemical analysis using the dissolution technique, the Co concentration could then match to the theoretical values.

CONCLUSIONS

We have developed a commercial scale production process for the synthesis of high quality WC/Co nanocomposite powders using readily available commercial water soluble W-, Co-, and C-containing chemical precursors. Different from all past literatures for the production of WC/Co composites, this process makes it possible for a full molecular mixing of all the required ingredients at the starting level in a large-scale. This methodology involves: 1) preparing aqueous solution of tungsten-, cobalt-, and carbon-containing compound precursor mixture; 2) spray conversion of the aqueous solution precursor mixture into a metalorganic precursor powder; 3) eliminating of the residual moistures or crystal water and organic ligands of N-H, and converting the C-H groups into a molecular mixed carbon source to obtain a pre-composite powder of W-Co-C-O system in a belt furnace under nitrogen at elevated temperatures; and 4) carburization of the precomposite powder in a rotary furnace under nitrogen at intermediate temperatures. Processing parameters have been developed and process controls have been established in each step of the material synthesis.

Analysis shows a high quality WC/Co nanocomposite with different compositions of Co that can be made, ranging from 0 wt% to 12 wt%. The carbon content of the nanocomposite can be well controlled to match industrial standard in the WC/Co hardmetal tool industry, e.g., for 88WC/12 Co and 94 WC/6 Co, the total carbon being ~5.4 ± 0.05 wt% for 94 WC/6 Co and ~5.8 ± 0.05 wt%, while the free carbon content being less than 0.1 wt% for both compositions. The WC phase is hexagonal and its grains size can be controlled to 30 - 80 nm range depending on processing parameters, while the Co binder phase has an fcc structure. The presence of this precipitates in the WC nanograins has also been confirmed by comparing the cobalt composition results obtained using different cobalt chemical analysis techniques.

The obtained WC/Co nanocomposite powders have also been consolidated into bulk parts by using a liquid phase sintering process. Preliminary studies have indicated that when grain growth inhibitors of Cr_3C_2 and VC are used, the WC grains in the consolidated parts are being ~200 - 300 nm average, with Co phase uniformly distributed in the WC grain boundaries. TEM studies have also illustrated a noticeable amount of nano-precipitate dispersion phase reside inside the WC nanograins. Preliminary µµd studies have identified that these precipitates are being the fcc cobalt phase.

Therefore, this paper mainly emphasizes the merits of a wet chemical approach using molecular mixing of W-, Co-, and C-containing water soluble precursors for the synthesis of commercial scale WC/Co nanocomposite powders, as well as some preliminary results of the consolidated bulk parts. Currently, we are investigating: 1) the detailed chemical reactions involved during each steps of the synthesis; 2) detailed microstructural analysis of the Co precipitates within the WC nanograins; and 3) extensive consolidation of WC/Co nanocomposite powders into high quality bulk parts.

ACKNOWLEDGEMENTS

This paper is dedicated to the memory of Mr. Dong Wei, who was the manager of the commercialization team of Fujian Jinxin Tungsten Co., Ltd. During the performance of powder synthesis, Mr. Dong Wei was a critical organizer in obtaining most of the synthesis and characterization data. The authors are highly grateful to the financial support of the China Ministry of Science and Technology under Contract No 2011DAE091304, as well as the financial and facility support from Fujian Jinxin Tungsten Co., Ltd. Particular thanks are also due to the following individuals at different facilities of Fujian Jinxin Tungsten Co., Ltd in order to perform experiments and obtain necessary data, including: Mr. Zhou Rongchang at the Spray Drying Processing Facility; Mr. Ni Ting at the Belt Furnace Processing Facility; Mr. Liu Benlong at the Rotary Furnace Process Facility, and Ms. Ge Sumei and Mr. Xiao Zhiwen at the Material Characterization and Testing Facility.

REFERENCES

1.   Moissan, H. (1897) The Electrical Furnace. French Edition, Translated by Lenher, V., Chemical Publishing Company, Revere.

2.   Voigtlander, H. and Lohmann, H. (1915) Metall-Fabrikations—G.m.b.H. German Patent 289,066.

3.   (1925) Patent-Treuhand-Gesellschaft fur elektriche Gluhlampen m.b.H. German Patent 420,689.

4.   Baumhauer, H. (1924) US Patent 1,512,191.

5.   Gortsema, F.P. (1976) US Patent 3,932,594, Union Carbide Corp.

6.   Gortsema, F.P. (1980) US Patent 4,190,439, Union Carbide Corp.

7.   Gortsema, F. and Kotval, P.S. (1976) Plansesee Seminar of Powder Metallurgy. Planseeberichte fur Pulvermetalurgie, 24, 254.

8.   Takatsu, S., et al. (1969) US Patent 3,440,035, Toshiba Tungalloy KK.

9.   Takatsu, S. (1971) Japanese Kokai 46/19300, Toshiba Tungalloy Co., Ltd.

10. Takatsu, S. (1978) A New Continuous Process for Production of WC-Co Mixed Powder by Rotary Kilns. Powder Metallurgy International, 10, 13.

11. Miyake, M. and Hara, A. (1979) On the Carbothermic Reduction of WO3 Powder in Nitrogen Atmosphere. Journal of the Japan Society of Powder and Powder Metallurgy, 26, 16-21. http://dx.doi.org/10.2497/jjspm.26.16

12. Hara, A., Miyake, M. and Yamamoto, T. (1975) Studies on Direct Carburization of WC form the Mixture If WO3 and Carbon. Journal of the Japan Society of Powder and Powder Metallurgy, 22, 12-16.http://dx.doi.org/10.2497/jjspm.22.12

13. Miyake, M., Hara, A. and Sho, T. (1979) Me-thod for Making Metallic Carbide Powders. Journal of the Japan Society of Powder and Powder Metallurgy, 26, 90-95.

14. Xiao, T.D., Zhang, Z.T. and Wang, D.M. (2009) US Patent No. 7,625,542.

15. Qiao, Y., Fischer, T.E. and Dent, A. (2003) The Effects of Fuel Chemistry and Feedstock Powder Structure on the Mechanical and Tribological Properties of HVOF Thermal-Sprayed WC-Co Coatings with Very Fine Structures. Surface and Coatings Technology, 172, 24-41.http://dx.doi.org/10.1016/S0257-8972(03)00242-1

16.   Guillemany, J.M., Dosta, S., Nin, J. and Miguel, J.R. (2005) Study of the Properties of WC-Co Nanostructured Coatings Sprayed by High-Velocity Oxyfuel. Journal of Thermal Spray Technology, 14, 405-413.

17.   Zucker, G., Downey, J., Bahr, D., Stephens, F. and Hager, J. (2002) Method for Production Tungsten Carbide. US Patent Application No. 20020009411.

18.   McCandlish, L.E., Kear, B.H. and Kim, B.K. (1990) Chemical Processing of Nanophase WC-Co Composite Powders. Materials Science and Technology, 6, 953-957.http://dx.doi.org/10.1179/mst.1990.6.10.953

19.   Kear, B.H. and McCandlish, L.E. (1993) Chemical Processing and Properties of Nanostructured WC-Co Materials. Nanostructured Materials,   3,   19-30.http://dx.doi.org/10.1016/0965-9773(93)90059-K

20.   Mohan, K. and Strutt, P.R. (1996) Microstructure of Spray Converted Nanostructured Tungsten Carbide-Cobalt Composite. Materials Science and Engineering: A, 209, 237-242.Mohan, K. and Strutt, P.R. (1996) Observation of Co Nanoparticle Dispersion in WC Nanograins in WC-Co Cermets Consolidated from Chemically Synthesized Powders. Nanostructured Materials, 7, 547-555. http://dx.doi.org/10.1016/0965-9773(96)00028-1

21.   Jia, K. and Fischer, T.E. (1996) Abrasion Resistance of Nanostructured and Conventional Cemented Carbides. Wear, 200, 206-214. http://dx.doi.org/10.1016/S0043-1648(96)07423-6.

22.   O'Yang, Y.F., Wu, Y.F. and Peng, Z.H. (1999) WC/Co Composite Powder Synthesized Using Fluidized Bed, and Its Applications. Chinese Tungsten Industry, 14, 220-215. Xu, T. (2009) The Quality Characteristics of WC/Co Nanocomposite Powders Produced by the Spray Conversion Process. In: Huang, B.Y. and Yi, J.H., Eds., 2009 Symposium of China Academic Meeting on Powder Metallurgy, 257-266.

23.   Shao, G.Q., Wu, B.L., Wei, M.K. and Yuan, R.Z. (1999) Development of WC Hardmetals with Ultrafine Grain Size. Journal of Wuhan University of Technology—Materials Science, 21, 1-5.

Comparative Study of the Adsorption and Desorption Behavior of Single and Multi-Ring Aromatics in Sediment Fractions

Chiedu N. Owabor[1], Samuel E. Agarry[2], Bamidele V. Ayodele[1], Ikechukwu S. Udeh[1], and Endurance Ehiosun[1]

[1]Department of Chemical Engineering, University of Benin, Benin City, Nigeria

[2]Department of Chemical Engineering, Ladoke Akintola University, Ogbomoso, Nigeria

ABSTRACT

The sorption behavior of benzene, toluene, ethyl benzene, xylene and naphthalene using clay and sand sediments under ambient conditions is examined in this study. Experimental results showed that, the time taken to attain adsorption equilibrium for naphthalene, and BTEX were 28, 30, 30, 32, 28 hrs and 20, 22, 22, 24, 22 hrs while the desorption equilibrium time were 10, 13, 12, 15, 12 hrs and 9, 9, 9, 11, 10 hrs in clay and sand respectively. All of the naphthalene, and BTEX were adsorbed at the different equilibrium times, using clay while the amount of naphthalene and BTEX adsorbed by sand, at different equilibrium times were 117, 121, 127, 123 and 134 mg. Following the results of the adsorption/desorption experiments, quantitative measurements showed that sand exhibited higher affinity for the solute as retained more chemicals (as high as between 58% - 66%) within it pores while nearly all the chemicals adsorbed by the clay were released at the attainment of equilibrium. The implication of this is that occlusion within the sand particles may likely be the resultant effect of continued sand-chemicals contact. The amount of contaminant solute adsorbed and desorbed affirmed that clay has a better capacity to retain naphthalene and BTEX than sand and this may not be unrelated to its large surface area, high porosity and higher hydraulic conductivity for the solutes arising from its good binding sites (small pore sizes) that tend to hold the adsorbates to its particles.

INTRODUCTION

The major environmental concern in urban and industrial areas can be attributed to the increasing level of pollution particularly by substances of organic origin. Various toxic chemicals such as polycyclic aromatic hydrocarbons (PAHs): naphthalene, anthracene, benzo(a)pyrene, phenanthrene, benzene, toluene, ethyl benzene and xylene (BTEX), heavy metals and dyes are continuously discharged into the environment as industrial waste, causing water, air and soil pollutions. These chemicals due to their recalcitrant persistent nature have relatively low solubilities in water, but are highly lipophilic. The four and more condensed aromatic rings are considered to be more dangerous than the two and three rings PAHs [1-3]. The presence of these compounds which are listed as priority pollutants [4-7] in the environment is of considerable public health and ecological concern due to their toxicity to a wide range of biological systems. These effects apart from the degradation of the ecosystem, also results in commodity loss, loss to the communities and economic loss due to spill cleanup cost.

Studies have shown that solute transport with linear equilibrium is an integral component of the degradation and/or mineralization of these toxic chemicals [8-12]. Contaminant transport is significantly viewed from two possible scenarios; fast sorption/desorption and slow sorption/desorption. Sorption tends to separate the direct contact between microorganisms and contaminants, which is necessary for biodegradation to occur. The practical effect of the adsorption and desorption rates, is that it controls the overall reaction rate of degradation process. They are one of the primary factors which affect availability, mobility and toxicity of contaminants in the soil [3-17]. They determine the measured concentration and the mechanism of distributing the contaminants into surfaces and into pores of individual soil particles [18-22] and are thus counteractive to efficient biodegradation.

Adsorption is a physical separation process in which certain compounds of a fluid phase are transferred to the surface of a solid adsorbent [23]. The separation is dependent on one component in a mixture being more readily adsorbed than the other components. The adsorption process takes place in three steps: macro transport, micro transport and sorption. Macro transport involves the movement of the organic material through the water to the liquid-solid interface by advection and diffusion. Micro transport involves the diffusion of the organic material through the macropore system of the soil particle to the adsorption sites in the micropores and submicropores of the soil particle. Adsorption results from the influence of Van der Waals forces which is essentially physical in nature. Due to the fact that the forces are not strong, the adsorption may be easily reversed. However, in some systems, additional forces bind adsorbed molecules to the solid surface. These are chemical in nature involving the exchange or sharing of electrons, or possibly molecules forming atoms or radicals. In such cases, the term chemisorption is used to describe the phenomenon [24]. This is less easily reversed than physical adsorption, and regeneration may be a problem.

The present study seeks to compare the processes of adsorption and desorption of naphthalene and BTEX in clay and sand as sediments with a view to determining the capacity and extent to which each sediment retains solutes at equilibrium as these primary processes form the basis for their availability for microbial uptake and mineralization. This is significant because in the long run results from the study will serve as a benchmark for the removal of these recalcitrant compounds which compromise the integrity of the environment. Suitable kinetic models which best describes the sorption-desorption mechanisms of the organic chemicals on the sediments will also be determined.

MATERIALS AND METHODS

Materials

The sediments used in this study were obtained from Ikpoba River, Benin City, Edo State, Nigeria. On collection, the sediments were stored in a black polythene bag. Pretreatment of the sediments was carried out via removal of stones and other heavy particles. A 220 µm mesh was used to remove the large non clay fractions from the clay and a 2.5 mm mesh was used to sieve the sand. Finally, they were then dried overnight at 383 K in a vacuum oven and stored in an air-tight container covered with a black polythene bag prior to the adsorption and desorption experiments described by [14].

The adsorbates, naphthalene, benzene, toluene, ethyl benzene and xylene were obtained from an auto-mechanic workshop at Ojota, Lagos state, Nigeria while the distilled water used for sample preparation, dilution and solution was obtained from the Department of Chemical Engineering, University of Benin, Benin City, and Edo State, Nigeria.

Methods

Adsorption-Desorption Kinetics Experiments

The rates of adsorption by the sediments (clay and sand) were determined from the uptake levels of PAHs and BTEX from aqueous solution in batch experiments before and after contact until adsorption equilibrium was attained in the clay and sand. 100 g of each of the sediments (clay and sand) were mixed with 500 ml of aqueous solution of the adsorbates at room temperature following the procedure of

Owabor et al. [25]. The slurry suspensions were sampled at intervals of 2 hours and UV spectrophotometer was used for sample analysis.

Adsorption Kinetics Models

The adsorption kinetics which describes the mechanism of the adsorption process in a given system were determined in this study using pseudo-first order model as defined by Lagergren and described by Chang et al. and Sivaraj et al. [14,25], pseudo-second order model [14], Elovich Model [26], intraparticle diffusion model [27] and the power function technique [28].

RESULTS AND DISCUSSION

The results of the experimentation and computation analysis of the kinetic modeling of the adsorption of naphthalene, benzene, toluene, ethylbenzene and xylene contaminated clay and sand are presented. The potential for the two sediments to adsorb and subsequently release the solutes for eventual mineralization by microbes have been exploited. The importance of this study was to explore the possible effects of sorption and bioavailability on biodegradation rates and the prediction of risk likely to occur from prolonged exposure of the soil to the contaminant solutes. The information on the mobility and hence availability of the chemicals is crucial in remediation studies as it determines the extent of partitioning and sequestration of the chemicals with sediments. A comparative analysis of the predictions from the kinetic modeling data obtained from experiments provided the basis for the choice of the applicable and suitable mechanisms for describing the adsorption and desorption processes.

Equilibrium Time for Adsorption

From Figure 1, the adsorption and desorption results showed that equilibrium was achieved at 28 and 20 hrs for naphthalene in the clay and sand fractions respectively. Similarly, for benzene, toluene, ethyl benzene and xylene, equilibrium was attained at 30, 30, 32 and 28 hrs and 22, 22, 24 and 22 hrs as shown in Figures 2-5 respectively. From the equilibrium time, sand was observed to reach equilibrium faster than clay for all of contaminant solutes used in this study. This can be attributed to its large pore spaces, intra-porous nature of sand (its pore spaces are connected to one another) and its higher permeability.

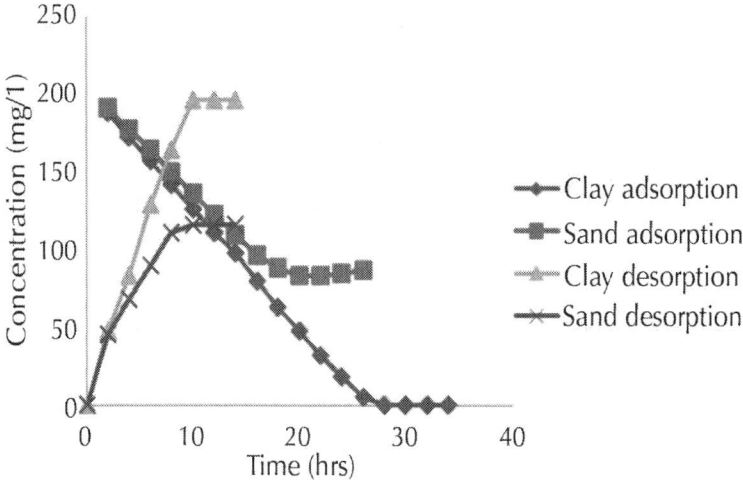

Figure 1: Variation of naphthalene concentration with time for adsorption and desorption.

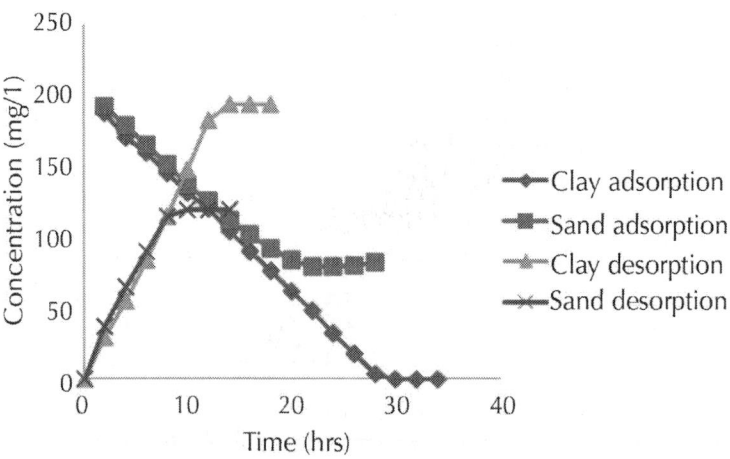

Figure 2: Variation of benzene concentration with time for adsorption and desorption.

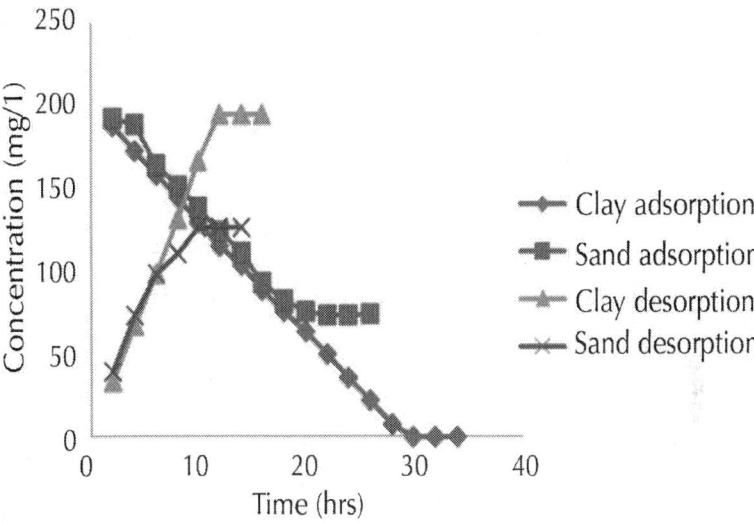

Figure 3: Variation of toluene concentration with time for adsorption and desorption.

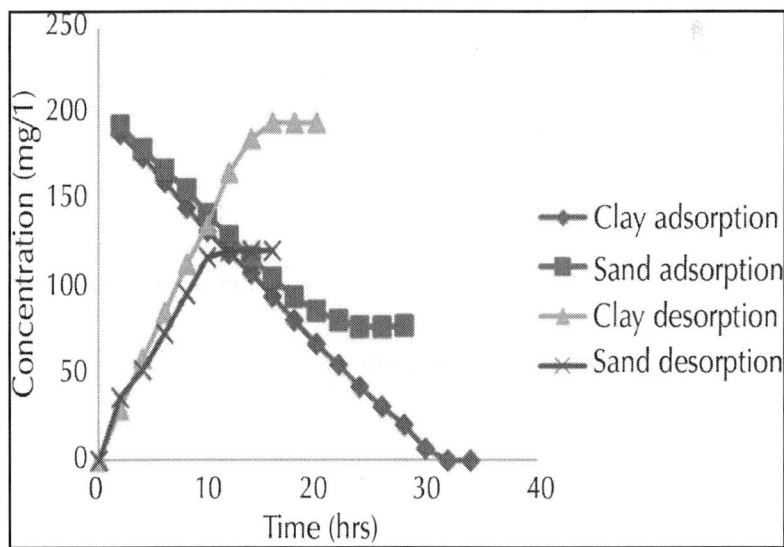

Figure 4: Variation of ethyl benzene concentration with time for adsorption and desorption.

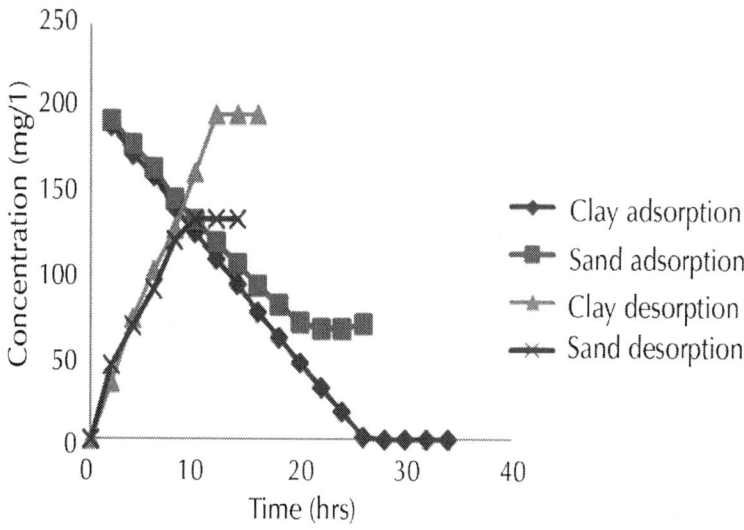

Figure 5: Variation of xylene concentration with time for adsorption and desorption.

However, it is worthy to note that before equilibrium was attained for both clay and sand, the amount of naphthalene adsorbed in the clay sample was 102 mg while in sand, the amount was 91 mg. Similarly, for benzene, toluene, ethyl benzene and xylene using clay, the amount adsorbed were 96, 97, 92 and 107 mg while in sand, the amount adsorbed were 88, 89, 83 and 95 mg. Based on the amount adsorbed given equal time before equilibrium, clay would be described as a better adsorbing agent of naphthalene and BTEX than sand because it holds more of naphthalene, benzene, toluene, ethyl benzene and xylene. This property of clay over sand can be attributed to its larger surface area and higher porosity than sand [29, 30]. Sand has low porosity though not as many pore spaces because its grains are very large such that in a unit of sand, the fraction of soil volume that consists of holes is a lot less than for clay soil. Clay has many small pore spaces in which water containing the contaminant solutes remains clinging to the clay particle surfaces. Porosity is an important consideration when evaluating the potential volume of water or amount of hydrocarbons sediments may contain. Sediments with higher porosity typically have higher hydraulic conductivity, a property of sediments that describes the ease with which water can move through pore spaces.

Equilibrium Time for Desorption

Desorption equilibrium, for clay sediment, was achieved at 10 hrs with 197 mg of naphthalene desorbed, while using sand, equilibrium was achieved at 9 hrs with 116 mg of naphthalene desorbed. For benzene, toluene, ethyl benzene and xylene, in clay, equilibrium was achieved at 13 hrs with 193 mg of benzene desorbed, 12 hrs with 194 mg of toluene desorbed, 15 hrs with 195 mg of ethyl benzene desorbed and 12 hrs with 195 mg of xylene desorbed respectively. While in sand, equilibrium was achieved at 9 hrs with 119 mg of benzene desorbed, 9 hrs with 126 mg of toluene desorbed 11 hrs with 121 mg of ethyl benzene desorbed and 10 hrs with 132 mg of xylene desorbed respectively.

Laboratory results obtained from this study affirmed that the desorption process was slow for both sediment types. The implication of this is that the organic chemicals are very slowly released for uptake or mineralization. The amount desorbed by the sediments may have been regarded as a result of their hydrophobicity. The solutes have a high affinity to sediments and great tendency to bind with organic carbon, mineral surfaces and interstitial voids within the micropores and submicropores of the sediment fraction [17,22,31], The observed slowly desorbing fraction can therefore be attributed to the effect of intraorganic matter and hindered pore diffusion mechanisms.

Mechanism of Adsorption and Desorption Processes

Tables 1-4 show the results of the kinetic modeling of the adsorption and desorption of naphthalene, benzene, toluene, ethyl benzene and xylene onto clay and sand used in this study. While the value of the reaction rate constant k predicted for the adsorption kinetics by the pseudo-first order and power function technique model equations closely approximated, there were however, wide deviations for the pseudo-second order, Elovich and intra-particle models. In contrast, for the desorption kinetics only the pseudo-first order and Elovich equations gave good fits as the intraparticle and power function equations were characterized by negative rate constants. The pseudo-second order rate constant was found to be ambiguous. However, estimations from the coefficient of regression indicate that the power function model best described the mechanism of adsorption of naphthalene, benzene, toluene, ethyl benzene and xylene while the intra-particle model gave the best description for desorption of naphthalene, benzene, toluene, ethyl benzene and xylene. They closely approximated to unity. Interestingly, the observed good approximations of the correlation coefficients obtained from the power function technique and intra-particle model for the contaminant solutes were consistent for both clay and sand sediments. This further affirms the suitability of the two models.

CONCLUSIONS

This study showed that adsorption and desorption of PAHs (naphthalene) and BTEX occurred in clay and sand with clay adsorbing more of the naphthalene, benzene, toluene, ethyl benzene and xylene than sand while sand desorbed more of the naphthalene, benzene, toluene, ethyl benzene and xylene than clay. The adsorptive property of clay over sand was attributed to its large surface area, higher porosity and high hydraulic conductivity of the adsorbates while the slow desorption can be attributed to its good binding sites (small pore sizes) that tend to hold the adsorbates to its particles. For both adsorption and desorption, equilibrium was attained faster for sand than for clay as a result of the higher permeability of sand sediment. The power function model gave the best description for the adsorption mechanism of naphthalene, benzene, toluene, ethyl benzene and xylene, while the intraparticle model was most suitable for describing the mechanism of desorption of the contaminant solutes.

Table 1: Kinetic parameters and correlation coefficient (R^2) values for the adsorption of naphthalene and BTEX using clay sediment

Kinetic Models	Parameters	Naphthalene	Benzene	Toluene	Ethyl benzene	Xylene
Pseudo-First Order	k_1	0.1210	0.116	0.136	0.095	0.150
	q_e	3.7547	3.8729	7.0781	3.2904	4.8211
	R^2	0.8202	0.7510	0.8440	0.8550	0.6842
	k_2	4×10^{-5}	9×10^{-5}	1.3×10^{-5}	1.259×10^{-5}	0.00242
Pseudo-Second Order	q_e	43.1034	27.7778	23.6407	-714.2857	30.8642
	R^2	0.0348	0.2087	0.3178	0.0002	0.056
	a	1.2458	1.2933	1.3063	1.3317	1.2430
Elovich		0.2723	0.2572	0.2581	0.2454	0.2775
	R^2	0.9179	0.8974	0.9075	0.9051	0.9241
	k_p	0.496	0.483	0.477	0.469	0.496

Intra-Particle	C	-0.737	-0.746	-0.723	-0.739	-0.718
	R^2	0.978	0.977	0.983	0.985	0.978
Power Function	v	1.013	0.981	0.981	1.022	1.009
	k	0.0684	0.0714	0.0684	0.0607	0.0702
	R^2	0.989	0.995	0.996	0.996	0.986

Table 2: Kinetic parameters and correlation coefficient (R^2) values for the adsorption of naphthalene and BTEX using sand sediment

Kinetic Models	Parameters	Naphthalene	Benzene	Toluene	Ethyl benzene	Xylene
Pseudo-First Order	k_1	0.1730	0.176	0.293	0.18	0.168
	q_e	2.1706	2.6013	1.6023	3.0465	2.7020
	R^2	0.9240	0.914	0.934	0.917	0.909
	k_2	0.0016	0.0003	0.0033	4.996×10^{-4}	2×10^{-5}
Pseudo-Second Order	q_c	6.3776	13.3333	-3.6657	-9.9404	-53.195
	R^2	0.1407	0.0436	0.1661	0.1045	0.0154
	A	2.419	1.9873	1.7778	1.9066	1.7794

Elovich	B	0.2054	0.2106	0.2120	0.2002	0.2325
	R^2	0.6067	0.9534	0.934	0.945	0.9472
	K_p	0.330	0.327	0.380	0.340	0.379
Intra-Particle	C	-0.389	-0.395	-0.543	-0.461	-0.488
	R^2	0.958	0.964	0.920	0.974	0.974
	V	1.008	1.019	1.176	1.081	1.045
Power Function	k	0.056	0.0522	0.0374	0.0428	0.0555
	R^2	0.966	0.968	0.969	0.979	0.979

Table 3: Kinetic parameters and correlation coefficient (R^2) values for the desorption of naphthalene and BTEX using clay sediment

Kinetic Models	Parameters	Naphthalene	Benzene	Toluene	Ethyl benzene	Xylene
Pseudo-First Order	k_1	0.275	0.207	0.255	0.207	0.202
	q_e	-1.033	-0.882	-0.695	-0.757	0.453
	R^2	0.922	0.852	0.796	0.877	0.893
	k_2	14.100	6.416	10.576	7.363	9.423

Pseudo-Second Order	q_e	0.018	0.035	0.029	0.027	0.025
	R^2	0.660	0.644	0.673	0.607	0.611
	A	2.558	2.646	2.604	2.688	2.688
Elovich	B	1.585	1.610	1.556	1.602	1.722
	R^2	0.937	0.883	0.904	0.912	0.913
	k_p	-0.387	-0.352	-0.363	-0.327	-0.354
Intra-Particle	C	1.300	1.379	1.351	1.335	1.327
	R^2	0.983	0.969	0.975	0.983	0.982
	V	-1.607	-1.240	-1.363	-1.302	-1.355
Power Function	k	0.979	0.977	0.993	1.112	0.989
	R^2	0.677	0.634	0.653	0.641	0.617

Table 4: Kinetic parameters and correlation coefficient (R^2) values for the desorption of naphthalene and BTEX using sand sediment

Kinetic Models	Parameters	Naphthalene	Benzene	Toluene	Ethyl benzene	Xylene
Pseudo-First Order	k_1	0.361	0.383	0.279	0.298	0.397
	q_e	1.533	-4.200	1.234	4.075	-1.382
	R^2	0.885	0.826	0.992	0.825	0.763
	k_2	35.270	34.908	33.385	26.800	31.607
Pseudo-Second Order	q_e	8.67×10^{-3}	8.67×10^{-3}	8.91×10^{-3}	9.43×10^{-3}	8.82×10^{-3}
	R^2	0.729	0.719	0.669	0.660	0.719
	a	5.076	4.329	4.237	4.484	4.065
Elovich		2.862	2.552	2.671	2.778	2.518
	R^2	0.948	0.958	0.974	0.921	0.948
	k_p	-0.200	-0.234	-0.237	-0.214	-0.241
Intra-Particle	C	0.627	0.740	0.740	0.751	0.797
	R^2	0.980	0.981	0.984	0.978	0.978
	V	-2.192	-1.800	-1.992	-1.626	-1.788
Power Function	k	0.600	0.429	0.624	0.549	0.610
	R^2	0.657	0.715	0.644	0.661	0.699

REFERENCES

1.   D. T. Sponza and R. Oztekin, "Destruction of Some More and Less Hydrophobic PAHs and Their Toxicities in a Petrochemical Industry Wastewater with Sonication in Turkey," Bioresource Technology, Vol. 101, No. 22, 2010, pp. 8639-8648. doi:10.1016/j.biortech.2010.06.124

2.   J. D. Stokes, G. I. Paton and K. T. Semple, "Behavior and Assessment of Bioavailability of Organic Contaminants in Soil: Relevance for Risk Assessment and Remediation," Soil Use and Management, Vol. 21, No. S2, 2006, pp. 475-486. doi:10.1079/SUM2005347

3.   W. Chen, L. Cong, H. Hu, P. Zhang, J. Li, Z. Feng, A. T. Kan and M. B. Tomson, "Release of Adsorbed Polycyclic Aromatic Hydrocarbons under Cosolvent Treatment: Implications for Availability and Fate," Environmental Toxicology and Chemistry, Vol. 27, No. 1, 2008, pp. 112-118. doi:10.1897/07-170.1

4.   J. Yan, L. Wang, P. P. Fu and H. Yu, "Photomutagenicity of 16 Polycyclic Aromatic Hydrocarbons from the US EPA Priority Pollutant List," Mutation Research/Genetic Toxicology and Environmental Mutagenesis, Vol. 557, No. 1, 2004, pp. 99-108. doi:10.1016/j.mrgentox.2003.10.004

5.   H. K. Bojes and P. G. Pope, "Characterization of EPA's 16 Priority Pollutant Polycyclic Aromatic Hydrocarbons (PAHs) in Tank Bottom Solids and Associated Contaminated Soils at Oil Exploration and Production Sites in Texas," Regulatory Toxicities and Pharmacology, Vol. 47, No. 3, 2007, pp. 288-295. doi:10.1016/j.yrtph.2006.11.007

6.   M. C. Bruzzoniti, M. Fungi and C. Sarzanini, "Determination of EPA's Priority Pollutant Polycyclic Aromatic Hydrocarbons in Drinking Waters by Solid Phase Extraction-HPLC," Analytical Methods, Vol. 21, No. 6, 2010, pp. 739-743. doi:10.1039/b9ay00203k

7.   S. Mitra and R. Pranab, "BTEX: A Serious Ground Water Contaminant," Research Journal of Environmental Sciences, Vol. 5, 2011, pp. 394-398. doi:10.3923/rjes.2011.394.398

8. E. F. Neuhauser, J. P. Kreitinger, D. V. Nakles, S. B. Hawthorne, F. G. Doherty, U. Ghosh, M. F. Khalil, R. S. Ghosh, M. T. O. Jonker and S. A. Van der Heijden, "Bioavailability and Toxicity of PAHs at MGP Sites," Land Contamination and Reclamation, Vol. 14, No. 2, pp. 261- 266. doi:10.2462/09670513.713

9. C. N. Owabor, S. E. Agarry and T. O. Azeez, "Development of a Transport Model for the Microbial Degradation of Polycyclic Aromatic Hydrocarbons in a Saturated Porous Medium," Journal of the Nigerian Association of Mathematical Physics, Vol. 16, No. 2, 2010, pp. 317-324.

10. W. Zhang, E. J. Bouwer and W. P. Ball, "Bioavailability of Hydrophobic Organic Contaminants: Effects and Implications of Sorption-Related Mass Transfer on Bioremediation," Ground Water Monitoring & Remediation, Vol. 18, No. 1, 1998, pp. 126-138.

11. C. N. Owabor, S. E. Ogbeide and A. A. Susu, "Adsorption and Desorption Kinetics of Naphthalene, Anthracene and Pyrene in Soil Matrix," Petroleum Science and Technology, Vol. 28, No. 5, 2010, pp. 504-514. doi:10.1080/10916460802108546

12. C. N. Owabor, S. E. Ogbeide and A. A. Susu, "Estimation of Transport and Degradation Parameters for Naphthalene and Anthracene: Influence of Mass Transfer on Kinetics," Environmental Monitoring and Assessment, Vol. 169, No. 1-4, 2010, pp. 607-617.doi:10.1007/s10661-009-1200-6

13. S. Y. Gebremariam, "Mineralization, Sorption and Desorption of Chlorpyrifos in Aquatic Sediments and Soils," Ph.D. Thesis, Washington State University, Pullman, 2011.

14. C. F. Chang, C. Y. Chang, K. H. Chen, W. T. Tsai, J. L. Shie and Y. H. Chen, "Adsorption of Naphthalene on Zeolite from Aqueous Solution," Journal of Colloid and Interface Science, Vol. 277, No. 1, 2004, pp. 29-34. doi:10.1016/j.jcis.2004.04.022

15. Z. Yu, W. Huang, J. Song, Y. Qian and P. Peng, "Sorption of Organic Pollutants by Marine Sediments: Implication for the Role of Particulate Organic Matter," Chemosphere, Vol. 65, No. 11, 2006, pp. 2493-2501. doi:10.1016/j.chemosphere.2006.04.036

16. R. S. Kookana, "The Role of Biochar in Modifying the Environmental Fate, Bioavailability and Efficacy of Pesticides in

Soils: A Review," Soil Research, Vol. 48, No. 7, 2010, pp. 627-637. doi:10.1071/SR10007

17.　C. N. Owabor and J. O. Osarumwense, "Pyrene Mineralization in Clay Soil with and without Organic Carbon: The Role of Adsorption and Desorption Kinetics Equilibria," Global Journal of Pure and Applied Sciences, Vol. 14, No. 1, 2008, pp. 109-113. doi:10.4314/gjpas.v14i1.16782

18.　H. M. Lesan and A. Bhandari, "Atrazine Sorption on Surface Soils: Time-Dependent Phase Distribution and Apparent Desorption Hysteresis," Water Research, Vol. 37, No. 7, 2003, pp. 1644-1654. doi:10.1016/S0043-1354(02)00497-9

19.　P. Wang and A. A. Keller, "Sorption and Desorption of Atrazine and Diuron onto Water Dispersible Soil Primary Size Fractions," Water Research, Vol. 43, No. 5, 2009, pp. 1448-1456. doi:10.1016/j.watres.2008.12.031

20.　W. J. Weber Jr. and J. C. Morris, "Kinetics of Adsorption on Carbon from Solution," Journal of the Sanitary Engineering Division, Vol. 89, No. 2, 1963, pp. 31-59.

21.　J. J. Pignatello and B. Xing, "Mechanisms of Slow Sorption of Organic Chemicals to Natural Particles," Environmental Science & Technology, Vol. 30, No. 1, 1996, pp. 1-11. doi:10.1021/es940683g

22.　W. Huang, "Effects of Organic Matter Heterogeneity on Sorption and Desorption of Organic Contaminants by Soils and Sediments," Applied Geochemistry, Vol. 18, No. 7, 2003, pp. 955-972.

23.　W. L. McCabe, J. C. Smith and P. Harriot, "Unit Operations of Chemical Engineering," 7th Edition, McGrawHill, New York, 2005, pp. 836-888.

24.　C. N. Owabor and S. E. Agarry, "Sorption Behaviour of Naphthalene in Clay and Coarse Sediments Kinetic and Equilibrium Studies," Nigerian Journal of Applied Science, Vol. 27, 2009, pp. 15-23.

25.　R. Sivaraj, C. Namasivayan and K. Kadirvelu, "Orange Peel as an Adsorbent in the Removal of Acid Violet 17 (Acid Dye) from Aqueous Solution," Water Management, Vol. 21, No. 1, 2001, pp. 105-110.

26.　M. N. Sahmoune, K. Louhab and A. Boukhiar, "Kinetic and Equilibrium Models for the Biosorption of Cr(III) on Streptomyces rimosus," Journal of Applied Sciences Research, Vol. 3, No. 4, 2008, pp. 294-301.

27.   W. J. Weber and T. M. Young, "A Distribute Reactivity Model for Sorption by Soils and Sediments. 6. Mechanistic Implications of Desorption under Supercritical Fluid Conditions," Environmental Science and Technology, Vol. 31, No. 6, 1997, pp. 1686-1691. doi:10.1021/es9605681

28.   S. Singh, L. K. Verma, S. S. Sambi and S. K. Sharma, "Adsorption Behavior of Ni(II) from Water onto Zeolite X: Kinetics and Equilibrium Studies," Proceedings of the World Congress on Engineering and Computer Science, 22-24 Octber 2008, San Francisco, pp. 112-117.

29.   C. W. Curry, R. H. Bennett, M. H. Hulbert, K. J. Curry and R. W. Faas, "Comparative Study of Sand Porosity and a Technique for Determining Porosity of Undisturbed Marine Sediment," Marine Georesource and Geotechnology, Vol. 22, No. 4, 2004, pp. 231-252.doi:10.1080/10641190490900844

30.   C. N. Owabor, I. O. Oboh and F. A. Omiojieahior, "Adsorption Isotherms for Naphthalene on Clay and Silt Soil Fractions: A Comparison of Linear and Nonlinear Methods," Advanced Materials Research, Vol. 367, 2012, pp. 359-364.doi:10.4028/www.scientific.net/AMR.367.359

31.   U. V. Okere and K. T. Semple, "Biodegradation of PAHs in 'Pristine' Soils from Different Climate Region," Journal of Bioremediation & Biodegradation, Vol. 3, 2012, p. 145.

Chapter 9

Enhanced Sorption of Naphthalene onto a Modified Clay Adsorbent: Effect of Acid, Base and Salt Modifications of Clay on Sorption Kinetics

Chiedu N. Owabor, Uzezi M. Ono, and Aigbokhan Isuekevbo

Department of Chemical Engineering, University of Benin, Benin City, Benin

ABSTRACT

This paper examined the influence of acid, base and salt modifications of clay on its rates of naphthalene adsorption. The modifiers used include hydrochloric acid (HCl), citric acid, sodium hydroxide (NaOH), ammonium hydroxide (NH₄OH), sodium chloride (NaCl) and zinc chloride (ZnCl$_2$). The results obtained showed that equilibrium adsorption of naphthalene from the bulk solution was attained at a faster rate using modified clay when compared with the unmodified clay. HCl-modified clay had the highest rate of adsorption with a surface area and porosity of 49.05 mm^2 and 53.4%. This was closely followed by NaOH-modified clay while down the order was the ZnCl$_2$-modified clay which had the least rate of adsorption with a surface area of 44.3 mm^2 and porosity of 43.4%. The implication of the retention time obtained from the equilibrium study is significant as it provides the bench mark for interplay between sorption and degradation for transport and transformation of contaminant solutes within the soil matrix.

INTRODUCTION

This research work relates to the generation of equilibrium data on the adsorption of naphthalene through experiment using modified and unmodified clay. The process of adsorption has been described as one of the most important chemical processes in soils. It determines the quantity of plant nutrients, metals, pesticides and other organic chemicals retained on soil surfaces and is therefore one of the primary processes that affects the transport of nutrients and contaminants in soil. Successful predictions of the transport and fate of solutes in the subsurface is hinged on the availability of accurate transport parameters. Data on availability, transport and fate of contaminant solutes, which are limited, can be obtained from sorption studies such as the one conducted in this study and hence be utilized for the design of reactors for effective treatment process.

The process of adsorption creates a film of the adsorbate (the molecules or atoms being accumulated) on the surface of the adsorbent. It differs from absorption, in which a fluid permeates or is dissolved by a liquid or solid. The term sorption encompasses both processes, while desorption is the reverse of adsorption. It is a surface phenomenon [1,2].

Various toxic chemicals such as polycyclic aromatic hydrocarbons, heavy metals, dyes, solvents have been discharged to the environment as industrial wastes, causing serious water, air and soil pollutions and they threaten the human health. Polycyclic aromatic compounds are continuously widely studied environmental subjects due to their potential carcinogenicity, mutageni-city or both. They are mainly emitted from combustion processes including engine exhaust, industrial processes, natural gas, domestic heating systems, barbecue, and smoke, incomplete combustion of fossil fuels, volcanic eruptions and forest fires [3-5]. The discharge of these toxic organic chemicals into water bodies causes harmful effects to the environment and treatment of the contaminated water as economically as possible has been a problem over time. However the use of sediment has become a viable alternative being naturally abundant, easily sourced and cheap.

The removal of these kinds of pollutants from the environment cannot be accomplished by using traditional methods. It is now extensively recognised that adsorption provides a feasible, effective method for the removal of pollutants from waste water [6, 7]. Research interests in the use of sediments such as clay, loamy soil, silt, shale etc. as adsorbent for the purpose of adsorption are currently being advanced [8-12]. The wide usefulness of these kinds of adsorbents is as a result of their high chemical and mechanical stability, and a variety of surface and structural properties. The pore structure and chemical properties generally determine the adsorption ability of the clays [13].

The natural form of clay shows relative ineffectiveness as an adsorbent for neutral organic contaminants such as polycyclic aromatic hydrocarbons. The adsorptive properties of sepiolite for neutral polycyclic aromatic hydrocarbons can be greatly improved by replacing the neutral inorganic interlayer cations (sodium and calcium ions) with large organic cations of the long chain alkyl hydrocarbons [14, 15]. These modified clays are called organoclays and have been effective in the removal of various neutral organic contaminants including naphthalene from aqueous solutions [16, 17]. Clay and other layered silicate clays are naturally hydrophilic. This makes them poorly suited to mixing and interacting with most polymer matrices which are mostly hydrophobic.

A popular and relatively easy method of modifying the clay surface, making it more compatible with an organic matrix, is ion exchanging. The cations are not strongly bound to the clay surface, so small molecule cations can replace the cations present in the clay. By exchanging it with various organic cations, montmorillonite clay can be compatible with a wide variety of matrix polymers [18]. At the same time, the process of modification helps to separate the clay platelets so that they can be more easily intercalated and exfoliated [19, 20].

This study therefore seeks to determine the role of acids, bases and salts on the performance of clay as an adsorbent in the removal of organic contaminants and at the same time, establish the adsorptive capabilities of both modified and unmodified clay.

MATERIALS AND METHODS

Materials

Clay used in this study was obtained from Ikpoba River, Benin City, Edo State, Nigeria. On collection, stones and other heavy particles were removed from the clay sample. It was then sieved through a 220 µm mesh size to remove the larger non-clay fractions from the clay. Clay obtained was shared into three portions. Each portion was modified with acid, base and salt.

Hydrochloric and citric acids, sodium and ammonium hydroxides, sodium and zinc chlorides used in this study were of analytical grade and obtained from the Chemical Engineering Laboratory, University of Benin, Benin City.

Methods

The clay samples (both modified and unmodified clay) were analyzed for the following parameters:

pH Determination

The soil pH was determined by using 10 g of clay sample into a 100 ml beaker and 20 ml of distilled water was added to the clay soil sample and stirred for 30 min. The sample pH was measured with the pH meter.

Bulk Density Determination

The bulk density of the clay was determined according to the procedure described by (Ahmedna et al., 1997; Huerta-Pujol et al., 2010). 100 g of clay was placed in a 100 ml measuring cylinder, a little vibration was applied until no particle space and constant level of clay was noticed in the cylinder and the volume recorded.

$$\text{Bulk density} \left(g/dm \right) = \frac{\text{Weight of clay}}{\text{Volume}}$$

(1)

Surface Area Determination

The surface area of the clay was determined by iodine adsorption method. The amount of aqueous solution was estimated by titrating a blank with standard thiosulphate solutions and compared with titrated values against iodine containing samples [21-23].

$$\text{Iodine value} \left(mg/g \right) = \frac{\left(B - S \right) \times M \times 126.91}{B \times W}$$

(2)

B and S are the volumes of thiosulphate solution required for blank and sample titrations respectively, W, the mass of activated carbon sample and M, the concentration (mol/l) of iodide solute and 126.91, atomic mass of iodine. The iodine value was used to measure the surface area.

Moisture Content Determination

A sample of wet clay was weighed, put into a Karl Kolb oven (Heraeus Type) to dry at a temperature of 95°C for one hour, cooled and then weighed following the procedure of Ramulu (2003).

$$\text{Percentage moisture} = \frac{\left(W_m - W_d \right)}{W_m} \times 100$$

(3)

where W_m and W_d are weights of sample and dry sample respectively.

Particle Density

20 g of oven dried soil was weighed into a measuring cylinder and after tapping the cylinder gently, the volume of the soil is recorded as V_1. 50 ml of distilled water was added slowly through the ride of the cylinder to soak the sample thoroughly. The final soil and water volume was noted as V_2, and the particle density calculated according to Equation (4) [24]:

$$\text{Particle density} = \frac{\text{Weight of soil } (20 \text{ g})}{\text{Volume of soil taken}} = \frac{W}{V_2 - V_1}$$

(4)

Modification of Clay Samples

Using Acid 200 ml of 0.4 M citric acid, and hydrochloric acid were added to two separate portions of 50 g prepared clay sample respectively. These were mixed in a magnetic stirrer for 3 hr at 200 rpm. The samples were centrifuged at 1500 rpm for 15 min. The samples were then rinsed with 500 ml of distilled water several times until they became neutral to litmus. The samples were then dried at 373 K in an oven [25, 26].

Using Base Two portions of the prepared clay sample were separately mixed in the ratio of 1 g to 5 ml of 0.4 M sodium hydroxide and ammonium hydroxide for 3 hr at 200 rpm. The samples were centrifuged at 1500 rpm for 15 min. The samples were then rinsed with 500 ml distilled water several times until they became neutral to litmus. The sample was then dried at 373 K in an oven [26-28].

Using Salt Two portions of the prepared clay sample were separately mixed in the ratio of 1 g to 7 ml 1M sodium chloride and zinc chloride at 303 K for 3 hr. The modified clay samples were then filtered and washed several times until they became neutral to litmus. Then the samples were dried at 373 K in the oven [29-31].

Preparation of the Bulk Solution

An aqueous solution of naphthalene was prepared by first dissolving 200 mg of naphthalene in 100 ml of ethanol before 900 ml of water was added. The solution was properly stirred to obtain a homogeneous solution.

Adsorption Studies

The rate of adsorption of the various modified clay samples were determined from the uptake levels of naphthalene from aqueous solution in batch experiments before and after contact until adsorption equilibrium was reached in the various modified clay adsorbent [32]. 100 g of modified clay were mixed with 1000 ml of aqueous solution of the adsorbate at room temperature. The slurry suspensions were sampled at intervals of 2 hr and UV spectrophotometer was used for sample analysis and the values of absorbance recorded.

RESULTS AND DISCUSSION

The results of the physico-chemical analysis of both the modified and unmodified natural clay samples are presented in Table 1.

The adsorption of naphthalene as shown in Figure 1 increased with increasing contact time and thereafter the adsorption from the aqueous phase remained constant with time. The patterns of the curve affirm that adsorption is time dependent. This result is consistent with literature reports of Gök et al. (2008), Owabor et al. (2010), and also agrees with the adsorption kinetic theory which states that the higher the time, the more the amount of fluid adsorbed in the adsorbent (McCabe et al., 1993).

Table 1: Physico-chemical analysis of modified and unmodified clay

Parameters	Unmodified	HCl modified	Citric acid modified	NaOH modified	NH$_4$OH modified	NaCl modified	ZnCl$_2$ modified
Bulk density (g/cm^3)	1.15	1.15	0.93	1.14	1.39	0.97	0.99
Surface area (mm^2)	46.52	50.05	48.15	48.10	46.00	47.35	44.30
Particle density (g/cm^3)	1.70	1.67	1.55	1.75	0.95	2.13	1.85
Moisture content (%)	4.25	4.1	4.15	4.1	4.21	3.95	3.85
pH	7.1	2.3	3.5	12.3	11.1	7.3	6.6
Porosity (%)	45.9	54.60	46.8	49.1	45.5	44.5	43.4

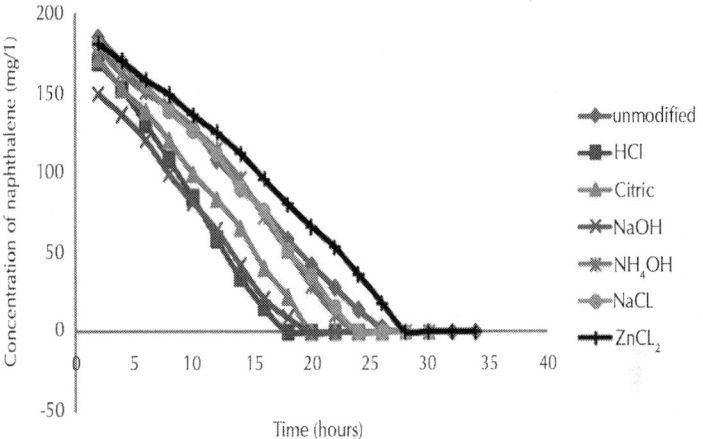

Figure 1: Variation of concentration with time for unmodified and modified clay samples.

Laboratory results showed that adsorption of naphthalene from the bulk solution using unmodified clay took the longest time of 28 hours before equilibrium was attained. Upon modification with the acids, bases and salts used in this study, the rate of adsorption increased an indication that equilibrium would be attained at a faster rate when compared with the unmodified clay.

The unmodified natural clay and modified clay samples were used as adsorbent for the adsorption of naphthalene. Both unmodified and modified clay samples adsorbed the adsorbate at different rates from the bulk solution under the same experimental conditions. The result obtained suggests that the hydrophilic nature of clay may be responsible for why it did not to a large extent adsorb the organic compound.

From this study, modifying with strong acidic solutions as shown using HCl significantly increased the adsorptive capacity of natural clay used as adsorbent. This was closely followed by the bases. The effect of citric acid was minimal on the sorption capacity of clay and this could be attributed to weakness of the acid. In contrast, the performance of the salts modified clay were very poor as virtually all the parameters determined in this study which are used to assess sorption capacity were notably below the values obtained for the unmodified natural clay. The implication of this result is that modification with salts adds little or no value to natural clay.

The rates of adsorption for the different modified samples of clay were studied using the concentration-time plots. The profiles are as shown in Figure 1.

The profiles showed that adsorption proceeded at a very slow rate for the unmodified samples. Here the rate of depletion of the adsorbate from the bulk solution took a long time and equilibrium was observed to be attained at the 28th hour.

Modifying with HCl showed a tremendous improvement in the adsorptive capacity of the clay. The rate of depletion of naphthalene from the bulk solution was observed to have increased and equilibrium attained at the 18th hr. This capacity can be adduced to the fact that HCl modified clay had the highest surface area and porosity; two very important factors that affect the rate of adsorption by an adsorbent. The interaction of the acid with the clay may likely have enhanced the pores of the sediment, thereby increasing its surface area and improving the adsorptive power of the clay.

The citric acid modified clay was a poorer adsorbent compared to HCl modified clay. Adsorption proceeded at a slower rate and equilibrium was attained at the 20th hour. This variation could be attributed to the strength of the acid involved as well as the decreased surface area and lower porosity obtained (Table 1).

Sodium hydroxide modified clay showed relatively good capacity to adsorb naphthalene from the bulk solution when compared to the performance of the HCl modified clay. It was observed that equilibrium was attained in the 20th hour. This time was the same as that observed for citric acid and thus suggests that a strong base was capable of chemically modifying clay just as a weak acid.

Ammonium hydroxide modified clay, as observed from Figure 1, equally showed a slow rate of adsorption of naphthalene from the bulk solution. It took a longer time precisely 24 hr to attain equilibrium.

Result in the case of sodium chloride modified clay showed that the rate of adsorption proceeded very slowly as compared with those of HCl and NaOH modified clay. As was observed with ammonium hydroxide, equilibrium was attained in 24 hours.

Zinc chloride modified clay presented in the profile of Figure 1 showed no difference in comparison with the unmodified natural clay. Equilibrium was reached at the same time observed for the unmodified clay i.e. 28 hours. Both sodium and zinc chlorides, showed comparatively slow rate of adsorption, which can be attributed to the low porosity and smaller surface area of the salt modified clay given in Table 1.

Generally, from the equilibrium time, sodium hydroxide compared favourably with citric acid and this can be explained from the standpoint of electro reactivity within the electrochemical series. Sodium hydroxide as a very strong base has a higher reactivity than a weak acid which in this case is represented by citric acid. This argument can again be advanced for the observed equilibrium time (28 hours) of zinc chloride and unmodified clay and ammonium hydroxide and sodium chloride (24 hours). Though ammonium hydroxide is a base and expectedly more corrosive than a salt, the equilibrium time from this study significantly showed that sodium chloride had a higher cation exchange capacity. Zinc is low in the electrochemical series and hence weak in electro positivity. It is again worthy to note that despite the fact that the molarity of the acids and alkalis were 0.4 M each, and the salts (NaCl and $ZnCl_2$) were 1.0 M, HCl showed the highest adsorption potential. This affirms that salts are poor modifiers. The effects of HCl, NaOH, NaCl and $ZnCl_2$ on the clay involved the exchange of their cations with those of the clay. This interaction predisposes the surface to adsorption. The physico-chemical analyses of the modified clay which gave a higher surface area and porosity when modified with HCl than those of the alkalis and salts, suggest that the cation exchange capacity of clay with HCl was better compared to NaOH and the salts.

The implication of modifying natural clay with HCl and NaOH is that it speeds up the hydraulic conductivity of the adsorbent for the adsorbate (naphthalene), an action which results from the fact that the small pore sizes tends to hold the adsorbates to its particles. Deductions from the physicochemical analysis further showed that salts reduced the moisture content and increased the maximum density and hence the particle density of the soil.

CONCLUSIONS

The adsorptive capacity of natural clay increased upon modification in the following order: salt < base < acid. Salts which are neutral to litmus did not show prospects as a modifying agent for clay. The dynamic behavior for the adsorption of naphthalene onto natural clay with various modifying agents increased to a large extent with increasing surface area and porosity of the modifier.

The obtained results indicate that modified clay, especially acid-modified clay can be a promising adsorbent for the removal of polycyclic aromatic hydrocarbons such as naphthalene from contaminated water.

REFERENCES

1.   J. F. Richardson, J. H. Harker and J. R. Backhurst, "Chemical Engineering: Particle Technology and Separation Processes," 5th Edition, Butterworth Heinemnn, India, 2003.

2.   W. L. McCabe, J. C. Smith and P. Harriot, "Unit Operations of Chemical Engineering," 7th Edition, McGraw Hill, New York, 2005.

3.   R. C. Zielke and T. J. Pinnavaia, "Toxicants: Binding of Chlorophenols to Pillared, Delaminated, and Hydroxy Interlayered Smectites," Clays and Clay Minerals Journal, Vol. 36, No. 5, 1988, pp. 403-408. doi:10.1346/CCMN.1988.0360504

4.   R. Goldman, L. Enewold, E. Pellizzari, J. B. Beach, E. D. Bowman, S. S. Krishnan and P. G. Shields, "Smoking Increases Carcinogenic Polycyclic Aromatic Hydrocarbons in Human Lung Tissue," Cancer Research Journal, Vol. 61, 2001, pp. 6367-6371.

5.   M. Ghiaci, A. Abbaspur, R. Kia and F. Seyedeyn-Azad, "Equilibrium Isotherms Studies for the Sorption of Benzene, Toluene and Phenol onto Organo-Zeolites and AsSynthesized MCM-41," Separation and Purification Technology, Vol. 40, No. 3, 2004, pp. 217-229. doi:10.1016/j.seppur.2004.03.001

6.   A. M. Mastral, T. Garcia, R. Murillo, M. S. Callen, M. V. Narvarro and J. Galban, "Assessment of Phenanthrene Removal from Hot Gas by Porous Carbons," Energy Fuels, Vol. 15, No. 1, 2001, pp. 1-7. doi:10.1021/ef000116g

7.   Ö. Gök, S. A. Özcan and A. Özcan, "Adsorption Kinetics of Naphthalene onto Organo-Sepiolite from Aqueous Solutions," Desalination, Vol. 220, No. 1-3, 2008, pp. 96- 107. doi:10.1016/j.desal.2007.01.025

8.   J. P. Chen, S. Wu and K. H. Chong, "Surface Modification of a Granular Activated Carbon by Citric Acid for Enhancement of Copper Adsorption," Carbon, Vol. 41, No. 10, 2003, pp. 1979-1986. doi:10.1016/S0008-6223(03)00197-0

9.   M. Khalid, G. Joly, A. Renaud and P. Magnoux, "Removal of Phenol from Water by Adsorption Using Zeolites Bain," Industrial & Engineering Chemistry Research, Vol. 43, No. 17, 2004, pp. 5275-5280. doi:10.1021/ie0400447

10.   A. S. Özcan and A. Özcan, "Adsorption of Acid Dyes from Aqueous Solutions onto Acid-Activated Bentonite," Journal of Colloid and Interface Science, Vol. 276, No. 1, 2004, pp. 39-46. doi:10.1016/j.jcis.2004.03.043

11.   A. Özcan and A. S. Özcan, "Adsorption of Acid Red 57 from Aqueous Solutions onto Surfactant-Modified Sepiolite," Journal of Hazardous Materials, Vol. 125, No. 1-3, 2005, pp. 252-259. doi:10.1016/j.jhazmat.2005.05.039

12.   W. T. Tsai, C. W. Lai and T. Y. Su, "Adsorption of Bisphenol-A from Aqueous Solution onto Minerals and Carbon Adsorbents," Journal Hazardous Materials, Vol. 134, No. 1-3, 2006, pp. 169-175. doi:10.1016/j.jhazmat.2005.10.055

13.   R. S. Juang, S. H. Lin and K. H. Tsao, "Mechanism of Sorption of Phenols from Aqueous Solutions onto Surfactant-Modified Montmorillonite," Journal of Colloid Interface Science, Vol. 254, No. 2, 2002, pp. 234-241. doi:10.1006/jcis.2002.8629

14.   S. Y. Lee and S. J. Kim, "Adsorption of Naphthalene by Hdtma Modified Kaolinite and Halloysite," Applied Clay Science, Vol. 22, No. 1-2, 2002, pp. 55-63. doi:10.1016/S0169-1317(02)00113-8

15.   C. F. Chang, C. Y. Chang, K. H. Chen, W. T. Tsai, J. L. Shie and Y. H. Chen, "Adsorption of Naphthalene on Zeolite from Aqueous Solution," Journal of Colloid and Interface Science, Vol. 277, No. 1, 2004, pp. 29-34. doi:10.1016/j.jcis.2004.04.022

16.   J. M. Hwu, G. J. Jiang, Z. M. Gao, W. Xie and W. P. Pan, "The Characterization of Organic Modified Clay and ClayFilled PMMA Nanocomposite," Journal of Applied Polymer Science, Vol. 83, No. 8, 2002, pp. 1702-1710.

17.   S. A. Boyd, M. M. Mortland and C. T. Chiou, "Sorption Characteristics of Organic Compounds on Hexadecyl Trimethy-

Ammonium-Smectite," Soil Science Society of America Journal, Vol. 52, No. 3, 1988, pp. 652-657.doi:10.2136/sssaj1988.03615995005200030010x

18.   N. Xu, C. Christodoulatos and W. Braida, "Adsorption of Molybdate and Tetrathiomolybdate onto Pyrite and Geothite: Effect of pH and Competitive Anions," Chemosphere, Vol. 62, No. 10, 2006, pp. 1726-1735.doi:10.1016/j.chemosphere.2005.06.025

19.   K. M. Spark and R. S. Swift, "Effect of Soil Composition and Dissolved Organic Matter on Pesticside Sorption," Science of the Total Environment, Vol. 298, No. 1-3, 2002, pp. 147-161.

20.   X. Wu, B. Xiao, R. Li, C. Wang, J. Huang and Z. Wang, "Mechanisms and Factors Affecting Sorption of Microcystins onto Natural Sediments," Environmental Science & Technology, Vol. 45, No. 7, 2011, pp. 2641-2647. doi:10.1021/es103729m

21.   M. Ahmedna, M. Johnson, S. J. Clarke, W. E. Marshall, and R. M. Reo, "Potential of Agricultural By-Products Based Activated Carbon for Use in Raw Surge Decolorization," Journal of the Science of Food and Agriculture, Vol. 75, No. 1, 1997, pp. 117-124.doi:10.1002/(SICI)1097-0010(199709)75:1<117::AID-JSFA850>3.0.CO;2-M

22.   O. Huerta-Pujol, M. Soliva, F. X. Martinez-Farre, J. Valero and M. Lopez, "Bulk Density Determination as a Single and Complementary Tool in Composting Process Control," Bioresource Technology, Vol. 101, No. 3, 2010, pp. 995-1001. doi:10.1016/j.biortech.2009.08.096

23.   J. A. Lori, A. O. Lawal and E. J. Ekaem, "Characterisation and Optimization of Deferration of Kankara Clay," Journal of Engineering and Applied Sciences, Vol. 2, No. 5, 2007, pp. 60-74.

24.   U. S. Sree Ramulu, "Principles in the Quantitative Analysis of Water, Fertilizers, Plant and Soils," Scientific Publishers, India, 2003.

25.   W. E. Marshall, L. H. Wartall, D. E. Boler, M. M. Johns, and C. A. Tolea, "Enhanced Metal Adsorption by Soyabeans Hulls Modified with Citric Acids," Bioresource Technology, Vol. 69, No. 3, 1999, pp. 263-268.

26. A. Gajo and B. Loret, "The Mechanics of Active Clays Circulated by Salts, Acids and Bases," 2007. http://www.unitn.it/ricerca/publicazioni.htm

27. P. Malakul, K. R. Srinivasan and H. Y. Wang, "Metal Toxicity Reduction in Naphthalene Biodegradation by Use of Metal-Chelating Adsorbents," American Society of Microbiology, Vol. 64, No. 11, 1998, pp. 4610-4613.

28. J. Q. Jiang and C. Cooper, "The Removal of Humic Substance with Modified Clay Adsorbents," Environ Engineering Science, Vol. 20, No. 6, 2003, pp. 581-586.doi:10.1089/109287503770736096

29. M. E. Chukwuedo and F. E. Okieimen, "Enhanced Metal Adsorption by Groundnut Husks Modified with Citric Acid," Journal of Chemical Society of Nigeria, Vol. 33, No. 2, 2008, pp. 50-53.

30. T. Abood, A. B. Kasa and Z. B. Chik, "Stabilization of Silt Clay Soil Using Chloride Compounds," Journal of Engineering Science & Technology, Vol. 2, No. 1, 2007, pp. 102-110.

31. M. F. Delbem, S. V. Ticiane, R. V. Francisco and R. D. Nicole, "Modification of a Brazilian Smectite Clay with Different Ammonium Salts," Química Nova, Vol. 33, No. 2, 2010, pp. 309-315. doi:10.1590/S0100-40422010000200015

32. C. N. Owabor, S. E. Ogbeide and A. A. Susu, "Adsorption and Desorption Kinetics of Naphthalene, Anthracene and Pyrene in Soil Matrix," Petroleum Science and Technology, Vol. 28, No. 5, 2010, pp. 504-514. doi:10.1080/10916460802108546

Citations

CHAPTER 1

Xiangwei Kong, Yuanhua Lin, and Yijie Qiu, "A New Method for Predicting the Position of Gas Influx Based on PRP in Drilling Operations," Journal of Applied Mathematics, vol. 2014, Article ID 969465, 12 pages, 2014. doi:10.1155/2014/969465.

CHAPTER 2

R. M. Miranda, J. Gandra and P. Vilaça (2013). Surface Modification by Friction Based Processes, Modern Surface Engineering Treatments, Dr. M. Aliofkhazraei (Ed.), ISBN: 978-953-51-1149-8, InTech, DOI: 10.5772/55986.

CHAPTER 3

Xiangwei Kong, Yuanhua Lin, Yijie Qiu, and Xing Qi, "A Novel Dynamic Model for Predicting Pressure Wave Velocity in Four-Phase Fluid Flowing along the Drilling Annulus," Mathematical Problems in Engineering, vol. 2015, Article ID 134102, 17 pages, 2015. doi:10.1155/2015/134102.

CHAPTER 4

M. Mahdavian, H. Atashi, M. Zivdar and M. Mousavi, "Simulation of CO2 and H2S Removal Using Methanol in Hollow Fiber Membrane Gas Absorber (HFMGA)," Advances in Chemical Engineering and Science, Vol. 2 No. 1, 2012, pp. 50-61. doi: 10.4236/aces.2012.21007.

CHAPTER 5

Jialin Tian, Chuanhong Fu, Lin Yang, et al., "The Wear Analysis Model of Drill Bit Cutting Element with Torsion Vibration," Advances in Mechanical Engineering, Article ID 254026, in press.

CHAPTER 6

Sun, K. , Bai, L. and Li, X. (2015) Analysis of the Chemical Safety Facility Investment Performance in China.Advances in Chemical Engineering and Science, 5, 102-109. doi: 10.4236/aces.2015.51011.

CHAPTER 7

Xiao, T. , Tan, X. , Yi, M. , Peng, S. , Peng, F. , Yang, J. and Dai, Y. (2014) Synthesis of Commercial-Scale Tungsten Carbide-Cobalt (WC/Co) Nanocomposite Using Aqueous Solutions of Tungsten (W), Cobalt (Co), and Carbon (C) Precursors. Journal of Materials Science and Chemical Engineering, 2, 1-15. doi: 10.4236/msce.2014.27001.

CHAPTER 8

C. Owabor, S. Agarry, B. Ayodele, I. Udeh and E. Ehiosun, "Comparative Study of the Adsorption and Desorption Behavior of Single and Multi-Ring Aromatics in Sediment Fractions," Advances in Chemical Engineering and Science, Vol. 3 No. 1, 2013, pp. 67-73. doi: 10.4236/aces.2013.31007.

CHAPTER 9

C. N. Owabor, U. M. Ono and A. Isuekevbo, "Enhanced Sorption of Naphthalene onto a Modified Clay Adsorbent: Effect of Acid, Base and Salt Modifications of Clay on Sorption Kinetics," Advances in Chemical Engineering and Science, Vol. 2 No. 3, 2012, pp. 330-335. doi: 10.4236/aces.2012.23038.

Index